Biodiversity, Conservation, and Environmental Management in the Great Lakes Basin

The Great Lakes Basin in North America holds more than 20 percent of the world's freshwater. Threats to habitats and biodiversity have economic, political, national security, and cultural implications and ramifications that cross the U.S.–Canadian border. This multidisciplinary book presents the latest research to demonstrate the interconnected nature of the challenges facing the Basin.

Chapters by U.S. and Canadian scholars and practitioners represent a wide range of natural science and social science fields, including environmental sciences, geography, political science, natural resources, mass communications, environmental history and communication, public health, and economics. The book covers threats from invasive species, industrial development, climate change, agricultural and chemical runoff, species extinction, habitat restoration, environmental disease, indigenous conservation efforts, citizen engagement, environmental regulation, and pollution. Overall the book provides political, cultural, economic, scientific, and social contexts for recognizing and addressing the environmental challenges faced by the Great Lakes Basin.

Eric Freedman is Knight Chair, Director of the Knight Center for Environmental Journalism, and Professor of Journalism at Michigan State University, USA. He has also won a Pulitzer Prize in Journalism.

Mark Neuzil is Professor of Communication and Journalism at University of St. Thomas, Minnesota, USA.

Biodiversity, Conservation, and Environmental Management in the Great Lakes Basin

Edited by Eric Freedman and Mark Neuzil

Routledge
Taylor & Francis Group
LONDON AND NEW YORK

earthscan
from Routledge

First published 2018 by Routledge

2 Park Square, Milton Park, Abingdon, Oxfordshire OX14 4RN
52 Vanderbilt Avenue, New York, NY 10017

Routledge is an imprint of the Taylor & Francis Group, an informa business

First issued in paperback 2019

British Library Cataloguing-in-Publication Data
A catalogue record for this book is available from the British Library

Library of Congress Cataloging-in-Publication Data
Names: Freedman, Eric, editor. | Neuzil, Mark, editor.
Title: Biodiversity, conservation and environmental management in the Great Lakes Basin / edited by Eric Freedman and Mark Neuzil.
Description: Milton Park, Abingdon, Oxon ; New York, NY : Routledge, 2018. | Includes bibliographical references and index.
Identifiers: LCCN 2017030705 | ISBN 9781138285811 (hardback) | ISBN 9781315268774 (ebook)
Subjects: LCSH: Great Lakes Region (North America)—Environmental conditions. | Biodiversity—Great Lakes Region (North America) | Nature conservation—Great Lakes Region (North America) | Environmental management—Great Lakes Region (North America) | Environmental policy—Great Lakes Region (North America) | Lake ecology—Great Lakes Region (North America)
Classification: LCC GE160.G75 B56 2018 | DDC 363.700977—dc23
LC record available at https://lccn.loc.gov/2017030705

ISBN: 978-1-138-28581-1 (hbk)
ISBN: 978-0-367-37699-4 (pbk)

Typeset in Goudy
by Apex CoVantage, LLC

Contents

Tables

Figures

Biographies

Editors

Eric Freedman Knight Chair, Director of the Knight Center for Environmental Journalism and Professor of Journalism at Michigan State University, where he is on the core faculty of the Canadian Studies Center and the Environmental Science and Policy Program. He is a Pulitzer Prize–winning journalist, and his books include *Environmental Crises in Central Asia: from Steppes to Seas, from Deserts to Glaciers* (Routledge) and *Great Lakes, Great National Forests: A Recreational Guide* (Thunder Bay).

Mark Neuzil Professor in the Department of Communication and Journalism at the University of St. Thomas in St. Paul, Minnesota. He is the author or coauthor of eight books, including *Environmental Crises in Central Asia: from Steppes to Seas, from Deserts to Glaciers* (Routledge) and has worked as a reporter and editor for the Associated Press, the *Minneapolis Star Tribune*, and other news organizations.

Contributors

Theresa Castor Associate Professor and Chair of the Communication Department at the University of Wisconsin–Parkside. She received her B.A.S. degree in rhetoric and communication and biochemistry, as well as her M.A. degree in rhetoric and communication, from the University of California, Davis, and her Ph.D. in speech communication from the University of Washington, Seattle. Her research focuses on discourse and language use in governance groups related to crises and environmental and weather issues.

Patricia Chow-Fraser Professor of Biology at McMaster University, Hamilton, Ontario, and teaches courses in applied ecology, environmental sustainability, and management of aquatic ecosystems. The Chow-Fraser lab uses remote sensing and GIS techniques, radio telemetry, and trophic-level manipulations to predict the effect of water level, impact of invasive species, and human disturbance on the quality and quantity of marsh habitat in the Great Lakes.

Tina Ciaramitaro Biological Science Technician with the U.S. Forest Service in Starkville, Mississippi, Forest Products Lab, Durability and Wood Protection Research. She assists with research on the control of eastern subterranean termites and wood-decay fungi. Previously, she worked with the Forest Service in Lansing, Michigan, Ecology and Management of Invasive Species and Forest Ecosystems, providing technical support for research on wood-boring insects. Her main focus was working with emerald ash borer.

Victoria M. Donovan Ph.D. student at the University of Nebraska, Lincoln. Her research interests include disturbance and landscape ecology, resilience and complex systems theory, and wildlife management. Her current research investigates the relationships between wildfire and system structure, resilience, and community dynamics.

Ran Duan Ph.D. student in the School of Journalism and Knight Center for Environmental Journalism at Michigan State University, where she earned her Master's in Journalism. She is pursuing a dual major in the Environmental Science and Policy Program. She received her B.A. in Journalism from Shandong University in China. Her research focuses on climate change visual communication, risk perception, and media coverage of environmental issues.

Marla R. Emery Research Geographer with the U.S. Forest Service, Northern Research Station, where her research focuses on the role of forests in providing food and sustaining cultures. She received her Ph.D. in Geography from Rutgers University. She has served as a Visiting Scholar at the College of Menominee Nation and is the National Coordinator of the Forest Service Native Peoples and Climate Change Research Program.

Jane Feely Master's student in Fisheries and Aquatic Biology at the University of Minnesota. She earned her bachelor's degree at the University of St. Thomas in St. Paul, Minnesota. Her research interests include natural resource management and wildlife conservation.

Jason L. Fischer Ph.D. student at the University of Toledo, Department of Environmental Sciences. His research focuses on evaluating biological responses to habitat restoration and structural integrity of restoration projects in large rivers.

David G. Flagel Assistant Director of the University of Notre Dame Environmental Research Center (West) and concurrent faculty member of the Department of Biological Sciences. He received his B.S. at the University of Wisconsin–Oshkosh and Ph.D. from the University of Notre Dame. His research focus is on trophic cascades involving mammals in terrestrial systems, particularly forests in the Great Lakes region of the United States.

Daniel Hayes Professor and Associate Department Chair at Michigan State University, Department of Fisheries and Wildlife. As part of the Partnership for Ecosystem Research and Management, his research focuses on cooperative research with the Michigan Department of Natural Resources on quantitative aspects of fish habitat and population dynamics. The effects of dams, dam removal, and invasive species have been a focal area of his work to improve the health of fish and fisheries.

B. Timothy Heinmiller Associate Professor in the Department of Political Science at Brock University where he researches and teaches in the areas of Canadian and comparative public policy. His research has been published in such journals as *Politics and Policy*, *Canadian Journal of Political Science*, *Natural Resources Journal*, *Review of Policy Research*, and *Governance*. His most recent book is *Water Policy Reform in Southern Alberta: An Advocacy Coalition Approach* (University of Toronto Press).

James F. Jenkins Deputy Clerk at the City of Sarnia and member of Walpole Island First Nation. He holds an M.A. in History from the University of Texas at Austin. He was previously employed at Walpole Island where he represented the First Nation on a multiagency watershed management strategy for the Thames River. He is currently involved in City of Sarnia initiatives to adopt and implement UNDRIP, as well as to review the governance structure of the city's environmental committees and its archaeological policies.

Apoorva Joshi Ph.D. student in Information and Media and Research Assistant for the Knight Center for Environmental Journalism at Michigan State University. She holds an M.A. in Environmental Science and Natural Resource Journalism from the University of Montana and a B.Sc. in Environmental Science from the University of Pune, India. She is pursuing a doctoral specialization in environmental science and policy and researches conservation criminology, environmental communication and behavior, and risk communication.

Gregory Kennedy Supervisory Fishery Biologist with the U.S. Geological Survey Great Lakes Science Center in Ann Arbor, Michigan. He works on physical habitat research in the Great Lakes using remote sensing techniques as it pertains to fish spawning. Most recently, he has been involved with numerous fish-spawning habitat restoration projects in the St. Clair River and Detroit River Areas of Concern.

Chantel Markle Ph.D. candidate at McMaster University in Hamilton, Ontario. She received her Honors B.Sc. from McMaster University with a specialization in biodiversity. Her Ph.D. research focuses on spatial ecology of freshwater turtles and use of remote sensing and geographic information systems for landscape-level conservation of at-risk turtles in Ontario, Canada. Current studies focus on mapping habitat suitability and occupancy for at-risk turtles, assessing strategies aimed at mitigating road mortality of reptiles, and impacts of invasive *Phragmites australis* on at-risk turtles.

Dalma Martinović-Weigelt Associate Professor and Director of the Environmental Science Program at the University of St. Thomas in St. Paul, Minnesota.

After completing her B.S. in Ecology at the University of Zagreb (Croatia), she pursued postgraduate education at the Central European University (Budapest) and the University of Mississippi. She received a Ph.D. in Fisheries Science and Water Resources at the University of Minnesota. Her research program focuses on characterizing the effects of environmental stressors and contaminants in aquatic ecosystems.

Deborah McGregor Associate Professor and Canada Research Chair, Indigenous Environmental Justice at York University in Ontario. Her Ph.D. is in Forestry from the University of Toronto. Her research has focused on indigenous knowledge systems and their various applications in diverse contexts, including water and environmental governance, environmental justice, forest policy and management, and sustainable development.

Robert McLaughlin Associate Professor and holder of a Partnership for Ecosystem Research and Management Scientist position between the University of Guelph and the Great Lakes Fishery Commission. He received his B.Sc. from the University of Windsor, M.Sc. from Queen's University, and Ph.D. from McGill University. His research program pursues curiosity-based and applied questions involving fish behavior, ecology, conservation, and management.

Matthew McLean Biological Science Technician for the U.S. Fish and Wildlife Service in Alpena, Michigan. His work focuses on monitoring for aquatic invasive species in the Great Lakes.

Andrea Olive Associate Professor of Political Science and Geography at the University of Toronto in Canada. She holds a M.A. in political science from Dalhousie University and a Ph.D. in political science from Purdue University. Her work focuses on wildlife and at-risk species conservation in Canada and the United States.

Lisa Peterson Ph.D. candidate, Quantitative Fisheries Center, Fisheries and Wildlife Department, Michigan State University. Her research interests include fisheries stock assessment, population dynamics modeling, management strategy evaluation, and structured decision-making. Her research is looking at methods for estimating natural mortality of fish populations using acoustic telemetry data. She also assists with the facilitation of a management strategy evaluation that engages stakeholders in the management of Lake Erie percid fisheries.

Angie Pigeon Director of Development for the Gun Lake Tribal Gaming Authority and board member for Gun Lake Investments, the tribe's economic development arm. She graduated cum laude from Western Michigan University with a Bachelor's of Business Administration and was instrumental in the development, construction, financing, and opening of the Gun Lake Casino. She belongs to the Heritage Award–winning Pigeon family known for its black ash basketry and has been part of Michigan State University Museum's Michigan Traditional Artist Apprenticeship Program.

Ed Pigeon Vice Chair of the Gun Lake Tribal Council. He has worked on such tribal projects as casino financing, land acquisition, and economic diversification and has lobbied on behalf of the Tribe at the state and federal levels. He belongs to Heritage Award–winning Pigeon family known for its black ash basketry and has many years of experience in cultural presentations and teaching activities, including language instruction to the Gun Lake Tribal community and integrating language components into cultural workshops.

Therese M. Poland Research Entomologist and Project Leader with the U.S. Forest Service, Northern Research Station, in Lansing, Michigan. She received her Master's degree in Pest Management and Ph.D. degree from Simon Fraser University in British Columbia, where she studied pheromone-based management of pine and spruce bark beetles. Her research is on biology, management, and regulatory issues related to invasive forest insect pests. She has worked on wood-boring species, including the pine shoot beetle, Asian longhorned beetle, and emerald ash borer.

Jesse N. Popp Adjunct Professor at Laurentian University and Wildlife Specialist with the Anishinabek/Ontario Fisheries Resource Centre. She holds an M.Sc. in Wildlife Ecology and a Ph.D. in boreal ecology. She is a wildlife ecologist with research interests that include wildlife management, restoration ecology, population ecology, spatial ecology, behavioral ecology, and the incorporation of traditional ecological knowledge in science.

Jeremy J. Pritt Fisheries Biologist for the Ohio Department of Natural Resources, Division of Wildlife Inland Fisheries Research Unit. He received a B.S. from the Ohio State University, an M.S. from Virginia Tech, and a Ph.D. from the University of Toledo. He conducts research on fish populations and fisheries management in reservoirs and rivers.

Nicholas J. Reo Assistant Professor of Native American and Environmental Studies at Dartmouth College. He studies indigenous knowledge and ecological stewardship on indigenous lands. His work blends ecological, anthropological, and indigenous methodologies in his work, often via tribal community–university partnerships.

Edward F. Roseman Research Fishery Biologist at the U.S. Geological Survey Great Lakes Science Center in Ann Arbor, Michigan. He specializes in fish early life history research and fish habitat restoration. For more than 10 years, he has led research and assessment programs to evaluate the restoration of fish spawning habitat in the Great Lakes, prey fish community dynamics in Lake Huron, and the life history ecology of native Great Lake's fishes such as walleye, lake whitefish, and lake sturgeon.

Brian Roth Associate Professor in the Department of Fisheries and Wildlife at Michigan State University. He conducts research on aquatic food web ecology and collaborates with the Michigan Department of Natural Resources as a member of the Partnership for Ecosystem Research and Management. His research interests include predator–prey dynamics, food web modeling, and the influence of non-native species in aquatic food webs.

Kathleen A. Rubio Former Research Assistant at the University of Michigan.

Heiko L. Schoenfuss Director of the St. Cloud State University Aquatic Toxicology Laboratory and a professor in the Department of Biological Sciences. His research focusses on the effects of contaminants of emerging concern on aquatic ecosystems. His research has identified links between treated wastewater effluent discharge and compromised fish health, between septic discharge and reduced reproduction in lake fish, and between storm water discharge and environmental contaminant loads. He has been the chair of the Society for Environmental Toxicology and Chemistry Endocrine Disruptor Testing and Risk Assessment Steering Committee and Advisory Committee and contributed expertise to the EPA Science Advisory Board.

Gaston Small Assistant Professor of Biology at University of St. Thomas in St. Paul, Minnesota. He received a Ph.D. in Ecology from the University of Georgia, where he studied nutrient dynamics in tropical stream ecosystems, and he completed postdoctoral training at the University of Minnesota, where he studied nitrogen cycling in the Great Lakes.

M.A. (Peggy) Smith Vice-Provost (Aboriginal Initiatives) and former Associate Professor in the Faculty of Natural Resources Management at Lakehead University in Thunder Bay, Ontario. Her Ph.D. in Forestry at the University of Toronto explored relationships around forest management between Indigenous peoples and the Province of Ontario. She continues her long-standing relationship as Senior Advisor to the National Aboriginal Forestry Association. Her current research focuses on free, prior, and informed consent of Indigenous peoples in natural resource development and conservation.

Bruno Takahashi Assistant Professor of Environmental Journalism and Communication in the School of Journalism and the Department of Communication at Michigan State University. He is also the research director of the Knight Center for Environmental Journalism. He received his B.A. in Communication from the University of Lima, Peru, and M.S. and Ph.D. degrees in Environmental Science from the State University of New York's College of Environmental Science and Forestry. He studies media coverage of environmental affairs, particularly in Latin America media.

Gail Vander Stoep Associate Professor in the Department of Community Sustainability, Michigan State University. She earned her bachelor's degree at the University of Texas at Austin and her master's and Ph.D. degrees at Texas A&M University. Her Ph.D. is in Recreation Resources Development, with foci on communications, interpretation, and social psychology. Her areas of specialization include interpretation of natural, cultural, and historic resources; use of interpretation and communications as a resource management tool; maritime heritage and coastal resources management; and sustainable tourism – all of which incorporate interpretive communication as a critical element.

Anthony Van Witsen Ph.D. student in the School of Journalism at Michigan State University. He studies the relationship between science and mass media with a focus on environmental science.

Kyle Powys Whyte Timmick Chair in the Humanities and Associate Professor of Philosophy and Community Sustainability at Michigan State University. He is a faculty member of the Environmental Philosophy and Ethics graduate concentration and a faculty affiliate of the American Indian and Indigenous Studies and Environmental Science and Policy programs. His research focuses on Indigenous peoples and climate justice, indigenous climate change planning, and the ethics of cooperation between Indigenous peoples and climate scientists.

Preface

Eric Freedman and Mark Neuzil

As the editors, the two of us approached this project from multiple vantage points. As scholars, we've researched how the media covers environmental issues in the Great Lakes Basin. As professors, we've taught environmental journalism and environmental communication to university students. As professional journalists, we've explained the latest scientific discoveries to the public and reported on topics ranging from air pollution to natural resource policy to public land conflicts to leaking landfills to endangered birds. As photographers, we've shot lighthouses, kayakers, wildlife, fish, rural cemeteries, vineyards, and sand dunes. As long-time residents of the Basin, we've canoed its rivers, hiked and cross-country skied its trails, swum at its beaches, biked its back roads, and camped at its parks. As authors, we've written books about its canoe heritage, its pioneering settlers, its forests, and its outdoor recreation opportunities.

We also saw a need for an in-depth look at a cross-section of key scientific, policy, and public administration research that will help shape the Great Lakes Basin's future as ecologically unique and economically vital. Thus this book provides multidisciplinary perspectives on an array of environmental, biodiversity, and conservation issues through the work and insights of U.S. and Canadian researchers, practitioners, policy-shapers, and scholars from a wide range of disciplines, including biology, fisheries and wildlife, forestry, chemistry, political science, economics, journalism and communication, community sustainability, and geography.

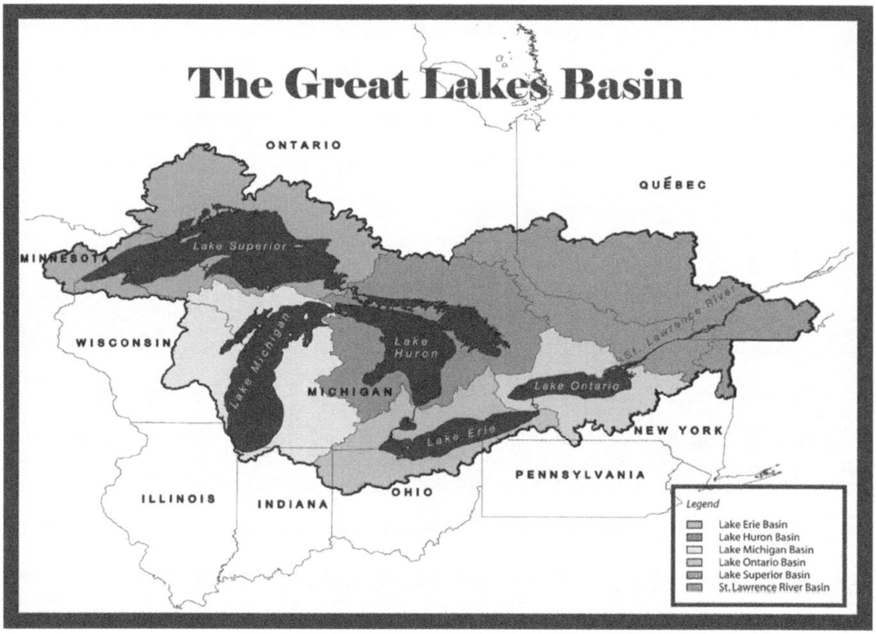

Figure 1.1 Map of the Great Lakes Basin.

Source: Council of Great Lakes Governors, inc.

1 Introduction

Examining the terrain

Eric Freedman and *Mark Neuzil*

The St. Clair River marks a 40-mile (64-kilometer) stretch of the U.S.–Canada border, connecting Lake St. Clair with the southern end of Lake Huron between Port Huron, Michigan, and Sarnia, Ontario. It boasts the largest freshwater delta in North America, shelters at least six shipwrecks, and supplies water for drinking and agriculture to communities on both sides of the river. It is home to 40 percent of Ontario's fish species and a major Great Lakes Basin wetland system. Geese fly low over the water. A riverside trail entices joggers, bicyclists, and dog-walkers. Beavers have returned. A shoreline restoration project on the Canadian side has improved habitat for aquatic wildlife. Pleasure boats cruise by. Bald eagles nest on a small island across from Guthrie Park in St. Clair Township, Ontario, watching to prey on fish and ducks.

But looks can be deceiving. Across from the park stands an oil refinery. Nearby is a cargo ship fueling station with high levels of mercury and polychlorinated biphenyls (PCBs) on the river bottom under its dock, contaminants that migrated downstream from a polluted industrial facility. The city of Sarnia is the location of "Chemical Valley," with its assemblage of petroleum refineries, chemical manufacturing facilities, and other heavy industries. Cargo vessels use the St. Clair River as a shipping channel. Under the Great Lakes Water Quality Agreement, the river has been designated as one of 43 Areas of Concern in the Great Lakes Basin since 1987 because of the history of industrial and urban uses along its shores. Thus several "beneficial uses" are classified as impaired, including eating fish and game, drinking water, and dredging. Scientists are still assessing other possible impairments, such as fish tumors, reproductive problems and deformities among birds and animals, and degraded wildlife and fish populations.

At the same time, there has been significant ecological progress along the river due to collaborative work by public agencies in both countries, local First Nations, environmental nongovernmental organizations, and industry, including heightened environmental compliance and regulation, removal of severely contaminated sediments, restoration of wildlife habitat, stronger spill prevention plans, and reduction of combined sewers that collect both sewage and stormwater runoff during heavy rainfalls.

We see a similar spectrum of environmental concerns in more remote, less-populated areas of the Basin. If you sail about 204 nautical miles (378 nautical kilometers) from Sarnia, Ontario, and Port Huron, Michigan, to the northern

Figure 1.2 Invasive phragmites threaten cattails and native grasses along the St. Clair River shoreline.

Source: Carin Tunney

end of Lake Huron, you'll arrive at Mackinac Island. Once a strategic military, fur trading, and missionary post, it is now a popular resort destination anchored in the Straits of Mackinac connecting Lake Huron and Lake Michigan as a single hydraulically linked system. Motor vehicles other than emergency vehicles are banned on the 3.8-square-mile (9.8-square-kilometer) island, so transportation for residents, visitors, and workers is by horse, bicycle, and foot.

Evidence of the island's history is long and deep, with prehistoric fishing camps found by archaeologists. Native American legend tells how the Great Spirit, Git-chi Man-i-tou, created the island (Gringhuis, 1970). European settlement followed. In 1695, French explorer-commander Antoine de la Mothe Cadillac wrote to a government minister about his ability to defend what was what New France called Michilimackinac Island – now Mackinac: "It is important that you be informed in case you are not, that this village is one of the strongest that there is in all of Canada. There is a fine palisaded fort, a highly disciplined garrison of elite soldiers" (Peyser and Brandão, 2008: 103). France, Britain, and eventually the fledgling United States vied for control of the island.

Development soon followed. By 1823, as Dunnigan (2008: 164) noted, "The island was crisscrossed with roads and trails . . . and visitors moved with relative ease as they toured natural and historical sites." By the mid-1820s, the U.S. Army established what is now the Mackinac Post Cemetery near Skull Cave, which was a burial site for Native Americans. Nineteenth-century geologists arrived to study the island's mineral potential. Its natural beauty attracted more and more visitors, such as a U.S. Army captain who described this scene in 1842: "This autumn time had clad the forests with a thousand hues . . . Between the crimson-leafed maple, the purple and gold ivy and beech, and the green eternal cedar . . . the naked rocks start out" (Dunnigan, 2008: 239). Vacation homes and hotels followed, as did souvenir stores, fudge shops, bars, bike rental and horse-drawn carriage businesses, and boutiques. Only about 500 residents live on the island year-round.

Although much of the island remains relatively pristine with heavy woods, geologically distinctive formations, and historic structures, the island and its environs in the Straits of Mackinac are not paradise. Despite stringent environmental protection laws, there is a risk that freighters and tankers passing through the straits may carry invasive species in their ballast water or may accidentally discharge fuel or other contaminants. The safety of the controversial Line 5 pipeline, commissioned in 1953 to carry crude oil and natural gas under the straits, is heatedly debated. The roughly 850,000 to 1 million visitors arriving at the island annually – most of them by high-speed ferry – put pressure on the infrastructure of the island and its gateway communities in Michigan's Upper and Lower Peninsulas. As a natural barrier, the straits isolate populations of wildlife, such as the bobcat, from the two peninsulas, thus impairing genetic diversity.

The setting

Superior. Michigan. Ontario. Erie. Huron.

Straddling the U.S.–Canadian border, the Great Lakes of North America – the Laurentian Great Lakes – are stunning, overwhelming, sprawling, awe-inspiring, vibrant, troubled, troubling, frustrating, threatening, threatened, enriching,

Figure 1.3 Ship passes under the Blue Water Bridge between Port Huron, Michigan, and Sarnia, Ontario.

Source: Pat Lambert

challenging, and reassuring. Ice Age legacies and symbols of the future. Repositories of incalculable natural wealth and targets of exploitation. Treasured for their cultural and economic value and abused through greed and ignorance. Sources of water essential for drinking, for farming, for industry, for transportation, for recreation, for shipping, for wildlife.

These lakes are the world's largest freshwater surface system. They dramatically influence climate. They provide water tainted with mercury, with PCBs. They offer invasion routes for hostile non-native species. Their storms pummel shoreline sand dunes and send ill-fated cargo ships and pleasure craft to the deep.

They are eyed with envy by thirsty parts of the continent. They are eyed with promise as sites for offshore wind farms. They are eyed with hunger for the wealth of their fisheries – even after years of heavy fishing pressure.

Encompassing the lakes themselves and covering far more territory, the Great Lakes Basin incorporates land in eight states and two provinces, covering roughly 297,000 square miles (770,00 square kilometers). Toronto, Windsor, Detroit, Toledo, Chicago, Buffalo, and Milwaukee are in the Basin, which spans 850 miles (1,400 kilometers) west–east and 725 miles (1,200 kilometers) north–south. Their dunes "wrought of glacial magic," as Elfont and Elfont (1997: 7) put it, are the largest that border freshwater in the world. An estimated 33 million people live in the Basin, alongside more than 75 other mammal species. "Which mammals occur here, where they live, and how they survive and reproduce is the result of a never-ending interplay between the living and nonliving components of the Great Lakes ecosystem" (Kurta, 1995: 3). The same is true, of course, of invertebrates, insects, birds, plants, and other flora and fauna.

Environmental and conservation challenges

As the St. Clair River and Straits of Mackinac examples illustrate, major ecological challenges confront the Great Lakes Basin. They include threats from invasive species, development and sprawl, climate change, agricultural and chemical runoff, air emissions, dependence on fossil fuels, energy exploration and hydraulic fracturing, wildlife and human diseases, poaching and illegal wildlife trafficking, and chemical contamination. These challenges to habitat and biodiversity have economic, political, national security, and cultural implications and ramifications that cross the U.S.–Canadian border.

We can trace the environmental impacts of settlements to the earliest days of cities along the Great Lakes and the St. Lawrence River, including Chicago, Montreal, Toronto, Rochester, Buffalo, Milwaukee, and Duluth. We can trace the environmental impacts of mega-projects that have irreversibly changed the very landscape, from the St. Lawrence Seaway to the Erie Canal to networks of superhighways. We have literally reversed the Chicago River's direction of flow.

Some species in the Basin have already gone extinct, including the eastern elk, the blackfin cisco, and most famously, the passenger pigeon. There are no precise figures on the number of critically endangered, endangered, and vulnerable

Figure 1.4 Boaters and spectators under the Blue Water Bridge where the St. Clair River meets Lake Huron.

Source: Dorothy Alexander

species remaining in parts or all of the Basin, but lists compiled by U.S. and Canadian government agencies are long. While a few are familiar iconic species, including the Canada lynx and the gray wolf, most are little-known to the general public, such as the copperbelly water snake, northern long-eared bat, Kirtland's warbler, Karner blue butterfly, dwarf lake iris, pitcher's thistle, and piping plover. The International Union for the Conservation of Nature (IUCN), which maintains the world's most comprehensive database of species' status, the IUCN Red List, cites reasons for that information gap, such as the fact that many species have not been fully evaluated.

These and other factors contributing to potential eco-devastation led a Lake Superior State University biologist to ask the audience at a local library where he was giving a public talk, "Are we at the beginning of the sixth mass extinction?" (Nolan, 2016).

Why should we worry about extinction? Doesn't it sound, well, abstract and distant, like the demise of dinosaurs and mastodons? Can't endangered species on the cusp of extinction be protected and bred in zoos and wildlife refuges? Can't science resurrect those that have already disappeared through cloning? What does it say about a society that lets species go extinct out of carelessness, callousness, avarice, or ignorance? The conservationist, philosopher, and wilderness advocate

Aldo Leopold once observed that humans could live comfortably without songbirds and wildflowers that lack "economic value" and whose disappearance and demise "would likely go unheralded." Leopold certainly would have agreed with environmental journalist Jason Mark's assessment that "species' instrumental value to humans shouldn't matter" because they have "an *intrinsic* value. They have a worth in and of themselves, even if it appears that they're good for nothing" (Mark, 2015: 85).

Can't government, businesses, science, and citizen action provide effective conservation management and environmental health protection for people, flora, and fauna in the Basin?

Overview of the book

The interplay of geography, hydrology, climate, biology, economics, politics, culture, history, and human emotion sets the stage for this book, which has three major aims:

- Illuminate the interconnected nature of habitat, biodiversity, and species extirpation and extinction challenges in the transborder Great Lakes Basin, which holds more than 20 percent of the world's freshwater.
- Highlight recent research into those challenges from multiple disciplines in the natural and social sciences.
- Provide political, cultural, economic, and social contexts for recognizing and addressing those problems and challenges.

We organized the book by four closely connected themes: (1) Habitat, Conservation, and Restoration; (2) Extinction and Survival; (3) Pollution, Climate Change, and Invasive Species; and (4) Public Policy.

Under Habitat, Conservation, and Restoration you'll read about the impact of dam removal projects on native and invasive fish species, the state of irrigation in the Great Lakes Basin, and the construction and restoration of reefs to provide safe harbors and enhance recreational fisheries. Chapters in the Extinction and Survival section explore efforts to protect habitat for at-risk turtle species, social attitudes toward the passenger pigeon as reflected by media coverage of the centennial of its extinction, the interaction of white-tailed deer and gray wolves, and patterns of movement by black bears and wolves. Chapters in the Pollution, Climate Change, and Invasive Species section highlight research on toxicants and chemicals of emerging concern; interactions among nutrient pollution, invasive species, and climate change; and efforts to save black ash as a Native American cultural resource from the destructive invasive emerald ash borer. The final theme, Public Policy, looks at U.S. congressional legislation about the Great Lakes, the role of conservation authorities in Ontario, public engagement and "citizen science" with the region's natural resources, and principles for environmental collaboration with Indigenous and non-Indigenous groups

Several threads connect these chapters, much like geology, hydrology, and geography connect the Great Lakes, the rivers flowing into them, and their watersheds. The first thread is the insignificance – at least to the forces of nature – of political borders between the two countries and among their states and provinces. Water flows, air moves, and wildlife travels easily across artificial lines on maps.

Another commonality is the complexity of the Basin's environmental and conservation management challenges from both a scientific and a public policy standpoint. There is no end in sight to the arrival of non-native flora and fauna that disrupt habitats and food supplies for native species. There is no end in sight to the creation of new chemicals whose short- and long-term health and environmental effects are unknown. There is no end in sight to population growth and to development pressures on fragile ecosystems and on limited natural resources. There is no end in sight to demands from water-hungry parts of the United States to "share" – in reality, to divert water from the Great Lakes. There is no end in sight to the threat of species extinction. And there is no end in sight to changing global climate that will impact lake levels, agricultural patterns, wildlife migration, human and animal health, economic activity, and outdoor recreational opportunities.

Borders do have political and diplomatic significance, however. What entities of government will take what legislative, regulatory, remedial, and protective steps to safeguard the region's natural resources? And who will pay for those measures? Thus another shared thread is the need for credible public and nongovernmental mechanisms and institutions to resolve environmental disputes. Researchers at Michigan Technological University and Northwest Indian College explained that a "multijurisdictional, multiscale system of governance" has evolved to address transborder problems of toxics such as mercury, PCBs, pesticides, and industrial chemicals that bioaccumulate in food chains. Scientists, they argue, can play important roles in connecting "various jurisdictional scales of governance" by monitoring atmospheric concentrations of such toxics and by "modeling their fate and transport" (Gorman et al., 2016: 443). Several such entities already exist, including the International Joint Commission and the Great Lakes Fishery Commission. In addition, environmental law enforcement and regulatory agencies collaborate and cooperate in investigations ranging from toxic discharges to trafficking in animals parts that cross state, province, and national borders.

Of course, it's impossible to cover all aspects of the Great Lakes Basin's environmental management, biodiversity, and habitat destruction and restoration in a single volume. That is why we, as editors, had to omit a number of relevant topics that also are the subject of fascinating research with potential practical impact, such as microplastics in the waters of the region, landfills and recycling capacity, mass transit systems, harbor dredging, zoning, and the role of nonprofit conservancies in protecting land and waters that government agencies cannot or will not protect.

Conclusion

This book underscores how complicated environmental issues are in the Great Lakes Basin, how difficult it is to make decisions based on a complex mix of science, economics, public opinion, and realistic policy alternatives – and how high the stakes are. "The one constant in the evolution of the Great Lakes since the arrival of Europeans is surprise," observed David Dempsey, a former environmental policy advisor for the International Joint Commission. He continued:

> No government and few individuals foresaw the collapse of fish stocks, the invasion of the sea lamprey, thousands of deaths from typhoid and cholera, the sudden fall of Lake Erie. And while individuals successfully pressed governments to undo these disasters, there was always a great cost. In the twenty-first century, the greatest cost may result from failing to act in advance of disaster.
>
> (Dempsey, 2004: 227)

References

Dempsey, D. (2004). *On the Brink: The Great Lakes in the 21st Century*. East Lansing, MI: Michigan State University Press.

Dunnigan, B. L. (2008). *A Picturesque Situation: Mackinac Before Photography 1615–1860*. Detroit: Wayne State University Press.

Elfont, C. J., and Elfont, E. A. (1997). *Sand Dunes of the Great Lakes*. Chelsea, MI: Sleeping Bear Press.

Gorman, H. S., Gagnon, V. S., and Norman, E. S. (2016). Local impacts, global sources: The governance of boundary-crossing chemicals. *History of Science*, 54(4): 443–459.

Gringhuis, D. (1970). *Lore of the Great Turtle: Indian Legends of Mackinac Retold*. Mackinac Island, MI: Mackinac Island State Park Commission.

Kurta, A. (1995). *Mammals of the Great Lakes Region*. Ann Arbor, MI: University of Michigan Press.

Mark, J. (2015). *Satellites in the High Country: Searching for the Wild in the Age of Man*. Washington, DC: Island Press.

Nolan, N. (2016, March 19). LSSU assistant professor examines extinction. *Sault News*, pp. 1, 3.

Peyser, J. L., and Brandão, J. A. (2008). *Edge of Empire: Documents of Michilimackinac, 1671-1716*. East Lansing, MI: Michigan State University Press

Part I

Habitat, conservation, and restoration

2 The dam dilemma for fisheries management in the Great Lakes

Daniel B. Hayes, Robert McLaughlin, Lisa Peterson, and Brian Roth

Introduction

About 179 native fish species occur in the Great Lakes (Roth et al., 2013). This level of species diversity is modest compared to other regions in the world, but Great Lakes fishes are important ecological resources in the world's largest freshwater ecosystem. They provide valuable sport and recreational fisheries. They are also important cultural icons for Native American and First Nation peoples, as well as for the broader cultural makeup of the Basin. Conserving the biodiversity of fishes in the region is important to maintain the benefits they provide, and the factors that contribute to overall species richness are important considerations in managing dams.

Although the word "biodiversity" is commonly used, a clear understanding of how ecologists and fishery biologists view this term is essential. The taxonomic composition of organisms in a given region is a key component of biodiversity (Noss, 1990). However, there are actually several components contributing to how biodiversity is measured and manifested in aquatic ecosystems. At the highest level, the total number of species present (e.g., the 179 native fish species) is often termed alpha (α) biodiversity. In the long run, α biodiversity increases when new species evolve from existing ones (speciation) or when new species immigrate into the Basin via natural processes (dispersal) or anthropogenic (human-caused) transportation. It decreases when species go extinct within the region. At the end of the last glaciation, the fish fauna in the Great Lakes was very limited. Over the past 11,000 to 14,000 years, many species have made their way into the Great Lakes Basin, and this process would be expected to continue, albeit slowly, in the absence of human intervention.

Alpha diversity can be subdivided further into beta (β) and gamma (γ) diversity. Beta diversity is the mean number of species present at a local site within the region. If there are many species present at each site within a large region, we expect the large region to have more species overall. Gamma diversity is the variation in the species among sites. This is commonly caused by site-to-site variation in habitat conditions. If a small number of species is found at each site, and the composition of those species differs widely from site to site, the overall α diversity is relatively high. Conversely, if a small number of species is found at each site, and the composition of those species is similar from site to site, the

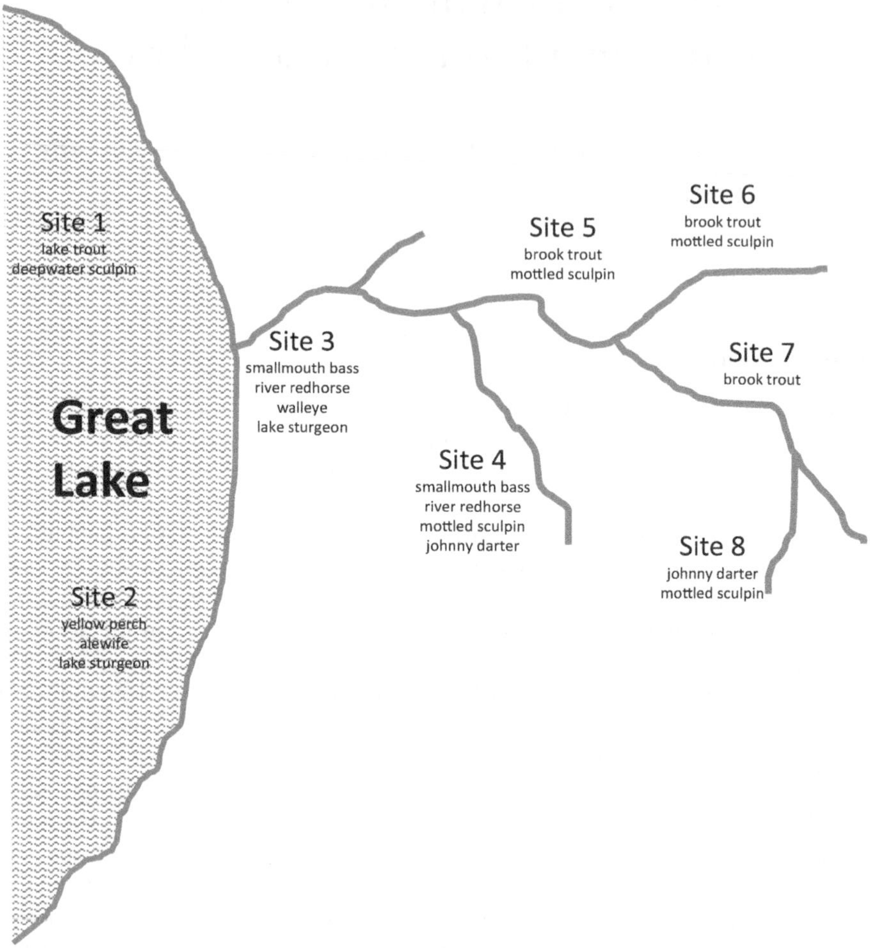

Figure 2.1 Hypothetical example of species that occur at different sites within an interconnected Great Lake and tributary river system.

Source: The authors

overall α diversity is low. The relationships among these forms of biodiversity are illustrated in Figure 2.1. In this hypothetical example, the overall α diversity is 11 species.

The mean number of species per site (β diversity) is 2.5 species. Gamma diversity has no direct measure, but we can see that the species composition of sites in the headwaters differs from sites near the river mouth, thereby contributing to the overall α diversity.

The fish faunal biodiversity of the Great Lakes region faces many threats, including invasive species, eutrophication, climate change, habitat loss, and degradation (Sala et al., 2000). We focus on the influences of dams on fish diversity in the Basin. Dams are widespread across the Basin, affecting nearly every major drainage and the types of fish occurring in each Great Lake. There are more than 12,000 dams in the Basin (Great Lakes Environmental Assessment and Mapping Project, 2016). Although the motivation for dam construction varies from site to site, each was built to provide societal benefits, such as power generation, flood control, maintenance of lake levels, and lumber and grain mills. Many of the dams in the region are old, having outlived their design lifespan, and no longer provide the benefits for which they were built (River Alliance of Wisconsin and Trout Unlimited, 2000; Doyle et al., 2003). The negative consequences of dam construction and operation on fish biodiversity have been increasingly recognized, but the interaction of dams with other processes affecting fish biodiversity, particularly invasions by non-native species, greatly complicate the situation (McLaughlin et al., 2013). Our goal is to provide an overview of how dams affect fish biodiversity and recommend a process for engaging in discussions regarding how dam construction, maintenance, and removal affect fishes in the Great Lakes.

Dams affect these populations in three general ways. First, they block the movement of fish between the Great Lakes and tributary streams, as well as within tributary streams. This often reduces or eliminates access to critical spawning and rearing habitats. Second, habitat conditions in impoundments are vastly different than in the original stream, and further, stream reaches below dams are typically changed relative to conditions prior to construction. Altered habitat conditions generally favor unassisted and assisted colonization of species not formerly found in the reach prior to construction. Third, changes in movement, habitat, and species composition affected by dams can alter the strength of interactions among fishes through introduction or loss of new species.

Research questions

To address our goal, we pose these research questions:

1. How much do dams diminish biodiversity in the Great Lakes region?
2. How much of this is due to habitat fragmentation versus habitat alteration?
3. What are likely impacts of climate change when added to dam effects?
4. How does blockage of invasive species play into this?

Method

In this chapter, we review and synthesize existing data and published studies.

Findings

Although there remains debate over losses and taxonomic status of some Great Lakes fishes, approximately 15 of the 179 fish considered native to the Basin have been extirpated (Cudmore-Vokey and Crossman, 2000; Roth et al., 2013; U.S. Fish and Wildlife Service, 2016). They are: paddlefish (*Polyodon spathula*), gravel chub (*Erimystax x-punctatus*), popeye shiner (*Notropis ariommus*), bigeye shiner (*Notropis boops*), weed shiner (*Notropis texanus*), harelip sucker (*Moxostoma lacerum*), deepwater cisco (*Coregonus johannae*), blackfin cisco (Coregonus nigripinnis), shortnose cisco (*Coregonus reighardi*), longjaw cisco (*Coregonus alpenae*), Arctic grayling (*Thymallus arcticus*), spoonhead sculpin (*Cottus ricei*), bluntnose darter (*Etheostoma chlorosomum*), gilt darter (*Percina evides*), and blue pike (*Stizostedion vitreum glacum*). Dams have not been proven to be a major contributor in the extirpation of any of these species, and as such, dams may not have had a direct impact on the α diversity of the Basin as a whole. However, dams have been identified as contributors to the loss of specific populations of fish, contributing to the risk of extirpation. For example, the Committee on the Status of Endangered Wildlife in Canada considers lake sturgeon (*Acipenser fulvescens*), American eel (*Anguilla rostrate*), and black redhorse (*Moxostoma duquesni*) imperiled in large part due to dams. It is also clear that dams can alter abundances, and even contribute to losses, of common species in sections of river above and below dams (Dodd et al., 2003; McLaughlin et al., 2006).

Fragmentation effects

Fragmentation of riverine habitats is one main way dams influence fish populations. Fragmentation affects species, such as lake sturgeon, that migrate from the Great Lakes to spawn in streams, as well as species that reside primarily in streams but move seasonally for spawning, such as brook trout (*Salvelinus fontinalis*). In both cases, fragmentation limits a population's ability to use habitats that provide critical resources for completion of the species' life cycle. For migratory fishes, the lowermost dam in a river's watershed typically has the greatest impact. For stream-resident fishes, it can be the cumulative effect of multiple dams or other barriers to movement, such as culverts, that reduces the population's ability to survive.

Drawing generalities about the fragmentation impact of dams is difficult. The impact depends on its size and location in the watershed, as well as the unique biological needs of the species using the river. It can also depend on the specific habitat conditions in a watershed. Lake sturgeon, however, provide an example of how fragmentation can adversely affect a species. Across the Basin, many streams that once supported lake sturgeon spawning runs have been dammed. As a consequence, recruitment has been drastically reduced and even eliminated from many systems. For example, of 24 lake sturgeon populations in Michigan waters, 12 are below the minimum viable population size (80 adults) and are at high risk of extirpation due to environmental and demographic stochasticity (Hayes and Carofinno, 2012). Four other populations are classified as small (80 to 200 adults) and at a high risk of declining to below the minimum viable size.

Low-head sea lamprey barriers also provide evidence that dams can fragment populations and facilitate species losses, with minimal effect on habitat conditions. In a comparison of 24 stream pairs (barrier and reference) with direct access to a Great Lake, the stream with a low-head sea lamprey barrier showed a net reduction of 2.4 fish species, on average, above the barrier. Streams with barriers also had four more species on average than those without barriers. Dodd et al. (2003) hypothesized that this difference was due in part to a concentration of fish below the barrier, thereby artificially increasing their vulnerability to capture and detection.

Specific examples of where river-resident fish populations are reduced by fragmentation caused by dams are uncommon. In a synthesis by Hayes et al. (2008), the authors found that naturalized populations of brown (*Salmo trutta*) and rainbow trout (*Oncorhynchus mykiss*) appeared to have little negative effect from fragmentation. However, numerous studies have provided evidence that movements of Great Lakes fishes within rivers, and between a river and a Great Lake, are more extensive than previously thought (Dolinsek et al., 2014). For example, Drake and Taylor (1996) found that movement of brook trout within Michigan's Ford River was extensive, likely due to marginally suitable temperatures in the lower reaches during summer. This suggests that fragmentation may interact with factors such as climate change or land cover changes that affect habitat suitability within river systems.

Habitat alteration effects

As with habitat fragmentation, it is difficult to generalize the extent of the impoundment created by dams and alterations to habitat conditions in downstream reaches. In particular, the impact of a dam on the thermal regime of a river depends on details specific to the situation. The height of the dam and volume of water impounded, the river discharge in relation to impoundment volume, as well as the surface area of the impoundment and exposure to sunlight all influence the amount of warming that occurs. In a study of 10 relatively large hydro dams in Michigan, Lessard and Hayes (2003) found temperature impacts ranging from none to a warming of 2.5 C°. Concurrent with warmer downstream temperature, Lessard and Hayes (2003) found substantial declines in brook and brown trout populations, with greater losses where temperature increases were largest.

Species interaction effects

Dams' effects on fishes resulting from altered species interactions is the least-studied aspect of dams. Lake-dwelling fish (lentic species) that colonize impoundments have the potential to compete with or prey upon stream-dwelling (lotic species) fishes. The walleye (*Sander vitreus*), largemouth bass (*Micropterus salmoides*), and common carp (*Cyprinus carpio*) are examples of species that may benefit from large impoundments but may also use tributary riverine habitats.

In contrast, native populations may benefit from the limitations that dams impose on non-native fishes (Rahel, 2013). The effects of sea lamprey (*Petromyzon marinus*) on native lake trout and whitefish have been limited by dams that deny lamprey access to spawning habitat. More than 70 low-head dams have been constructed in the Great Lakes Basin specifically to limit sea lamprey reproduction (Great Lakes Fishery Commission 2000). It is difficult to gauge the full effect of sea lamprey reduction on the biodiversity of the Great Lakes, but these reductions are viewed as crucial in allowing lake trout (*Salvelinus namaycush*) restoration, as well as reducing adverse effects on other native and non-native fishes such as lake sturgeon, lake whitefish (*Coregonus clupeiformis*), chinook salmon (*Oncorhynchus tshawytscha*), and coho salmon (*Oncorhynchus kisutch*).

Dams throughout the Basin also limit the upstream range of other species, such as the round goby (*Neogobius melanostomus*) and ruffe (*Gymnocephalus cernuus*) (Kornis and Vander Zanden, 2010). Likewise, dams are likely to limit the distribution of potential invaders, such as bighead carp (*Hypophthalmichthys nobilis*), silver carp (*Hypophthalmichthys molitrix*), and black carp (*Mylopharyngodon piceus*) if they eventually colonize the Great Lakes.

Combined effects of dams on fishes

The ideal "experiment" to determine the overall effects of dams would be to evaluate a fish assemblage before and after construction over sufficient time for the assemblage to respond fully. No studies on Great Lakes tributaries meet this ideal. Two alternative lines of evidence provide insight into the combined effects of dams on fish biodiversity, but with limitations. Comparisons of assemblages above and below existing dams can provide insight into how dams affect the distribution of fishes in the tributary, but such comparisons miss species that already were extirpated throughout the tributary following construction. Similarly, tracking the fish community above and below a dam prior to and after removal can provide causal evidence for changes in abundance and distribution. Both forms of comparison will also miss benefits and costs that extend beyond the watershed and into the lake, such as sea lamprey control on lake trout.

The Hayes et al. (2008) study most clearly demonstrated the effect of dams on assemblages by evaluating the degree of similarity between stream reaches with different levels of impacts by dams. In free-flowing stream reaches, the Sorenson's similarity index, which indicates the degree of similarity in species lists, averaged 0.69, providing a baseline expectation for "natural" systems. Morisita's similarity index, which indicates the degree of similarity in relative abundance of species in the assemblage, averaged 0.75 in these free-flowing streams. Both measures relate to γ diversity in that they relate to the degree of overlap in species composition among sites. As such, larger values of these indices generally indicate lower values of γ diversity. In streams with low-head barriers and streams with hydro dams that had little habitat effect on streams, the Sorenson index averaged 0.61 and 0.76, respectively, and the Morisita index averaged 0.52 and 0.87. Thus, these streams show little difference from the pattern in free-flowing streams. A set of streams with hydro dams, in contrast, showed a Sorenson index of 0.49 and a

Morisita index of 0.37. Both measures indicate a much greater degree of disparity in the assemblage, and hence impact on that assemblage, for rivers with both fragmentation effects and habitat alterations.

In an analysis of more than 10,000 stream sites in Wisconsin, Cumming (2004) found that although dams had an effect via fragmentation, their influence on habitat conditions, particularly discharge (volume of water in the stream) and summer maximum temperature, had a much greater influence on the sites' species richness (β diversity).

Case studies of removals in the Great Lakes Basin are relatively uncommon. The earliest study by Kanehl et al. (1997) reported responses of only selected individual species and aggregate measures of the assemblage such as the Index of Biotic Integrity (IBI). As such, it is difficult to put those findings within the framework we use in this chapter. Catalano et al. (2007) similarly used the IBI as an indicator of assemblage structure. However, their detailed analysis regarding the distribution of a suite of common species showed that nine found almost exclusively below the lower-most dam on Wisconsin's Baraboo River had expanded their distribution upstream following removal of four low-head dams.

Burroughs (2007) evaluated the impact of the removal of the Stronach Dam on the fish assemblage of the Pine River in Michigan. This dam was upstream of another dam, so the river reach evaluated lacked access for fishes resident in Lake Michigan. Prior to removal of the dam, the Morisita index between the river reach below the dam and above the impoundment was only 0.11, indicating a low degree of similarity. Following removal, the Morisita index increased to 0.80, a level similar to that of free-flowing streams. Also telling about the impact of the dam on fish distribution, of 35 species caught in total, 18 were found only below the dam prior to removal. Seventeen of the 18 were found upstream of the dam site following removal. These results show that although dams may increase γ diversity, as indicated by the similarity indices, this increase is artificial and caused by a loss of species above dams, which reduces β diversity.

Projected impacts of climate change

To date, the impacts of climate change are only beginning to manifest themselves, and as such, the projected impact on the fish diversity of the Great Lakes depends critically on the types and extent of changes in the climate regime that actually occur. We note, however, that fragmentation by dams limits the movements of fish and their ability to respond to habitat alterations caused by climate change. Changes to the climate regime will also likely interact with dams to change their impact on habitat conditions. For example, climate change is anticipated to increase the frequency of rare flood events. If dams are inundated more frequently, especially during the spring migration period, that could reduce the effectiveness of low-head dams in limiting the reproduction of invasive fishes.

Discussion and conclusion

Although the science for the Great Lakes is limited, sufficient theoretical reasons and empirical evidence suggest that dams will affect α, β, and γ diversity. There is little evidence that dams have reduced α diversity for the Great Lakes Basin as a whole, but they certainly have influenced β and γ diversity. Dams have reduced the mean number of species (β diversity) in many reaches above and below dams by fragmentation of habitats, leading to population reduction or loss. In the long run, this may lead to a loss of α diversity at the Great Lakes or Great Lakes Basin level. Changes in γ diversity due to dams are more complex. Dams tend to decrease the natural similarity in the fish community of streams by creating a discontinuity in habitat accessibility, as well as creating greater habitat disparity above and below dams. On its face, this leads to an increase in γ diversity, but this increase often coincides with a loss of β diversity. More research into the interplay of these different faces of diversity would help to better understand the suite of processes that lead to the overall diversity of the region.

When dams are built, the owner expects that the benefits that accrue will outweigh construction and maintenance costs. It is important to note, however, that environmental externalities, such as reductions in fish populations, generally do not – or historically did not – weigh into these costs (Orr et al., 2004). Thus, there is often an inherent conflict between the benefits a dam owner receives and the cost that society bears. More pointedly, as reservoirs fill in, as they inevitably do, benefits to the owner typically diminish, and maintenance costs increase, but environmental impacts generally remain. At some point, the cost–benefit ratio becomes unfavorable to the owner, and difficult choices are needed on whether to upgrade the dam and/or impoundment or remove the structure. Some owners simply abandon a dam because there are no financial benefits to removing it. As such, the ownership status of many dams is uncertain and may rest with a state, provincial, or federal government by default. Thus, there is a "legacy" of dams that provide little to no societal benefits.

Moving forward

It is clear that dams have a variety of far-reaching effects on the biodiversity of Great Lakes fish fauna and that there are trade-offs involved when thinking about modifying or removing a dam. Thus, decisions regarding the eventual fate of dams need to be made in a systematic and transparent way. The question is how to do that. We suggest that structured decision-making (SDM) provides an approach to dealing with different considerations in a clear and transparent manner.

The SDM process consists of a number of steps. Involving stakeholders at each step is crucial to the effectiveness of the process and support for the final decision. That becomes increasingly important when dam management decisions are controversial. Several key components are present in nearly all applications of SDM, although the specifics may vary from one situation to the next (Arvai et al., 2001; Clemen and Reilly, 2001; Gregory and Long, 2009; Gregory et al., 2012; Irwin et al., 2011). First, the decision-maker defines the context, specifying the problem and the situation that surrounds it. A second step involves specifying objectives and performance measures. An objective identifies the end state sought, while the performance measures quantify the success in achieving the objectives. Third, potential management actions that could satisfy the objectives are identified. By taking time to consider multiple approaches, decision-makers have a better chance of discovering a strategy that fulfills the objectives while minimizing trade-offs. Fourth, the consequences and uncertainties of each strategy are evaluated and compared. By being explicit about these consequences, the trade-offs of alternative strategies will become apparent, and the uncertainty can be used to think about risk. Different tools can be used to analyze these options, either qualitatively or quantitatively. With this information, decision-makers are better prepared to make a decision that is transparent, well thought out, and defensible.

SDM is not a new approach to natural resource management. It has been followed in water-use planning in British Columbia (Gregory and Failing, 2002), and SDM themes are found throughout the federal Canadian Wild Salmon Policy (DFO, 2005). In the United States, SDM was included in the technical guide for the conduct of adaptive management issued by the U.S. Department of the Interior (Williams et al., 2009). It has also been applied to a specific dam removal in the Great Lakes (Jones, 2014) and advocated for fisheries management in general (Irwin et al., 2011). Resource managers are recognizing SDM as a way to make transparent decisions about complex problems, which makes it well-suited for decisions about dams.

By applying SDM, decisions regarding the management of specific dams should include a road map leading to the final decision. This road map should be responsive to stakeholders' views without sacrificing the expert opinion of scientists and the experiences of managers. This decision-making process is flexible and, depending on the dam and the information at hand, could involve simply sketching out the problem or more rigorous and involved quantitative modeling. This process also provides a way to consider decisions on a broader scale, such as evaluating trade-offs involved with removing one dam and not another, or removing all the dams in a river system. SDM clarifies the context and issues surrounding a dam no matter its scale, reduces the bias inherent in a "gut" decision or one influenced by controversy, and results in consistent, coherent, and defendable decisions.

Using SDM to analyze the effects of dams enables us to take actions that maximize the biodiversity of the Great Lakes while taking into account the societal benefits and barriers to invasive species that dams also provide. To protect the diverse species and complex ecosystem services the Great Lakes provide, we must make decisions that can be supported by all the myriad of affected stakeholders. In the face of controversial issues, structured decision making can ensure everyone's voice is heard and everyone can see the reasoning behind a final decision.

References

Arvai, J. L., Gregory, R., and McDaniels, T. L. (2001). Testing a structured decision approach: Value-focused thinking for deliberative risk communication. *Risk Analysis*, *21*: 1065–1076.

Burroughs, B.A. (2007). *Effects of dam removal on fluvial geomorphology and fish*. Ph.D. Dissertation, Michigan State University, East Lansing, MI.

Catalano, M. J., Bozek, M. A., and Pellett, T. D. (2007). Effects of dam removal on fish assemblage structure and spatial distributions in the Baraboo River, Wisconsin. *North American Journal of Fisheries Management*, *27*: 519–530. doi: 10.1577/M06–001.1

Clemen, R., and Reilly, T. (2001). *Making Hard Decisions*. Belmont, CA: Duxbury.

Cudmore-Vokey, B., and Crossman, E. J. (2000). Checklists of the fish fauna of the Laurentian Great Lakes and their connecting channels. *Canadian Manuscript Report of Fisheries and Aquatic Sciences*, 2550. Available at https://pdfs.semanticscholar.org/c5bb/d5e7fbedd1232a15a97c34666a5c6d34cf48.pdf

Cumming, G. S. (2004). The impact of low-head dams on fish species richness in Wisconsin, USA. *Ecological Applications*, *14*: 1495–1506.

DFO (Fisheries and Oceans Canada). (2005). *Canada's Policy for Conservation of Wild Pacific Salmon*. Vancouver: DFO.

Dodd, H. R., Hayes, D. B., Baylis, J. R., Carl, L. M., Goldstein, J. D., McLaughlin, R. L., Noakes, D. L. G., Porto, L. M., and Jones, M. L. (2003). Low-head lamprey barrier effects on stream habitat and fish communities in the Great Lakes Basin. *Journal of Great Lakes Research*, *29*(Supplement 1): 386–402.

Dolinsek, I. J., McLaughlin, R. L., Grant, J. W. A., O'Connor, L. M., Pratt, T. C., and Magnan, P. (2014). Do natural history data predict the movement ecology of fishes in Lake Ontario streams? *Canadian Journal of Fisheries and Aquatic Sciences*, *71*: 1171–1185.

Doyle, M. W., Stanley, E. H., Harbor, J. M., and Grant, G. S. (2003). Dam removal in the United States: Emerging needs for science and policy. *EOS Trans AGU*, 84(4): 29–36. doi: 10.1029/2003EO040001

Drake, M. T., and Taylor, W. W. (1996). Influence of spring and summer water temperature on brook charr, "Salvelinus fontinalis", growth and age structure in the Ford River, Michigan. *Environmental Biology of Fishes*, *45*: 41–51. doi: 10.1007/BF00000626

Great Lakes Environmental Assessment and Mapping Project. (2016). *Tributary dams*. Available at http://greatlakesmapping.org/great lake stressors/6/tributary-dams

Great Lakes Fishery Commission. (2000). Sea lamprey barriers: Limiting sea lamprey access to spawning habitat. Available at www.glfc.org/pubs/FACT_5.pdf

Gregory, R., and Failing, L. (2002). Using decision analysis to encourage sound deliberation: Water use planning in British Columbia, Canada. *Journal of Policy Analysis and Management*, *21*: 492–499.

Gregory, R., Failing, L., Harstone, M., Long, G., McDaniels, T., and Ohlson, D. (2012). *Structured Decision Making: A Practical Guide to Environmental Management Choices*. West Sussex, UK: Wiley-Blackwell.

Gregory, R., and Long, G. (2009). Using structured decision making to help implement a precautionary approach to endangered species management. *Risk Analysis*, *29*: 518–532.

Hayes, D. B., and Carofinno, D. C. (2012). Michigan's lake sturgeon rehabilitation strategy. *Fisheries Special Report*, 62, Michigan Department of Natural Resources Fisheries Division, Lansing, Michigan.

Hayes, D. B., Dodd, H. R., and Lessard, J. L. (2008). Conservation considerations for small dams on cold-water streams. In: Nielsen, J. L., Dodson, J. J., Friedland, K., Hamon, T. R., Musick, J. and Verspoor, E. (Eds.), *Reconciling Fisheries with Conservation: Proceedings of the Fourth World Fisheries Congress*. Bethesda, MD: American Fisheries Society, Symposium 49, 1791–1806.

Irwin, B. J., Wilberg, M. J., Jones, M. L., and Bence, J. R. (2011). Applying structured decision making to recreational fisheries management. *Fisheries*, *36*(3): 113–122.

Jones, M. L. (2014). *Black Sturgeon River Dam Structured Decision Making Project*. Final Report to the Ontario Ministry of Natural Resources and Great Lakes Fishery Commission.

Kanehl, P. D., Lyons, J., and Nelson, J. E. (1997). Changes in the habitat and fish community of the Milwaukee River, Wisconsin, following removal of the Woolen Mills Dam. *North American Journal of Fisheries Management*, *17*: 387–400. doi: 10.1577/1548- 8675(1997)017<0387:CITHAF>2.3.CO;2

Kornis, M. S., and Vander Zanden, M. J. (2010). Forecasting the distribution of the invasive round goby (Neogobius melanostomus) in Wisconsin tributaries to Lake Michigan. *Canadian Journal of Fisheries and Aquatic Sciences*, *67*: 553–562.

Lessard, J. L., and Hayes, D. B. (2003). Effects of elevated water temperature on fish and macroinvertebrate communities below small dams. *River Research and Applications*, *19*: 721–732.

McLaughlin, R. L., Porto, L. M., Noakes, D. L. G., Baylis, J. R., Carl, L. M., Dodd, H. R., Goldstein, J. D., Hayes, D. B., and Randall, R. G. (2006). Effects of low-head barriers on stream fishes: Taxonomic affiliations and morphological correlates of sensitive species. *Canadian Journal of Fisheries and Aquatic Sciences*, *63*: 766–779.

McLaughlin, R. L., Smyth, E. R. B., Castro-Santos, T., Jones, M. L., Koops, M. A., Pratt, T. C., and Vélez-Espino, L.-A. (2013). Unintended consequences and trade-offs of fish passage. *Fish and Fisheries*, *14*: 580–604.

Noss, R. F. (1990). Can we maintain biological and ecological integrity? *Conservation Biology*, *4*: 241–243.

Orr, C., Roth, B., Forshay, K., Gonzales, J. D., Papenfus, M. M., and Wassell, R. D. G. (2004). Examination of physical and regulatory variables leading to small dam removal in Wisconsin. *Environmental Management*, *33*: 99–109. doi: 10.1007/s00267-003-2941-6

Rahel, F. J. (2013). Intentional fragmentation as a management strategy in aquatic systems. *Bioscience*, *63*: 362–372.

River Alliance of Wisconsin and Trout Unlimited. (2000). *A Citizen's Guide to Restoring Rivers*. Trout Unlimited, Arlington, VA.

Roth, B. M., Mandrak, N. E., Peters, J., Hrabik, T. R., and Sass, G. G. (2013). Fishes and decapod crustaceans of the Great Lakes Basin. In: Taylor, W. W. and Lynch, A. (Eds.), *Great Lakes Policy and Management*, 2nd edition. East Lansing, MI: Michigan State University Press, 105–135.

Sala, O. E., Chapin III, F. S., Armesto, J. J., Berlow, E., Bloomfield, J., Dirzo, R., Huber-Sanwald, E., Huenneke, L. F., Jackson, R. B., Kinzig, A., Leemans, R., Lodge, D. M., Mooney, H. A., Oesterheld, M., Poff, N. L., Sykes, M. T., Walker, B. H., Walker, M., and Wall, D. H. (2000). Global biodiversity scenarios for the year 2100. *Science*, *287*: 1770–1774.

U.S. Fish and Wildlife Service. (2016). *Extinct Species*. Available at www.fws.gov/midwest/endangered/lists/extinct.html

Williams, B. K., Szaro, R. C., and Shapiro, D. D. (2009). *Adaptive Management: The U.S. Department of the Interior Technical Guide*. Adaptive Management Working Group, U.S. Department of the Interior, Washington, DC.

3 Irrigation in the Great Lakes Basin

Prospects and conflicts

B. Timothy Heinmiller

Introduction

Irrigation is responsible for about 40 percent of global food production but most agriculture in the Great Lakes Basin has been rain-fed and, historically, irrigation has played a marginal role. Thus, the Great Lakes Basin has been largely spared the serious water conflicts that have characterized large-scale irrigation in other parts of the world.

Irrigation can play a crucial role in growing high-value crops and improving crop yields, but it is also a heavy consumer of water and can be a major contributor to water pollution, creating water management challenges and water use conflicts. With the onset of climate change, there is good reason to believe that irrigation might significantly increase in the Great Lakes Basin, creating the potential for water conflicts that have rarely been seen before. This chapter investigates irrigation and irrigation-related water conflicts in the Basin in the context of the changing climate. That includes the role of irrigation in Basin agriculture and recent trends in irrigation water use.

Using the conflict over Michigan's Mud Creek Irrigation District in the early 1990s, the chapter analyzes the prevailing institutions that will manage future irrigation-related water conflicts. Given that irrigation water use has grown substantially in recent years, irrigation has clear potential to create conflicts in the Basin, although recent institutional changes should provide an effective platform for containing and managing these conflicts.

Statement of the problem

Though often overshadowed by the region's industrial and service sectors, agriculture is an essential and broad-based economic activity in the Great Lakes Basin. Its deep, fertile soils and temperate climate combine to make the southern part an agricultural "sweet spot." Altogether, in the eight Great Lakes states and Ontario, agriculture involves about 114 million acres (45.6 million hectares) and about US$150 billion in annual production (Kling et al., 2003).

Unlike many other parts of the world – and even many other parts of North America – rain-fed agriculture is the norm in the Basin. Its climate is such that farmers can usually rely on spring and summer rains to provide the water necessary for their crops, such as corn, soybeans, and most types of grains. Prolonged droughts and dramatic floods have been relatively rare, at least thus far. A substantial amount of irrigation is present, but only as a supplement to rainfed agriculture. Irrigation is usually used to ensure that drought-sensitive crops get appropriate amounts of water at crucial times, but most farmers still rely on precipitation as the primary means of watering their crops (Annin, 2006). The large-scale irrigation systems prevalent in the western United States and Canada, involving massive dams, diversions, and canal systems, are absent in the Great Lakes Basin. Instead, most irrigation is small-scale, with farmers using pump-and-sprinkler systems to irrigate from local streams and groundwater sources. No irrigation water comes from inter-Basin diversions or dam storage, and little is pumped directly from the Great Lakes.

Depending on the year, irrigation is the second- or third-largest consumptive water use – a use that removes water and returns very little of it – in the Basin, accounting for 16 to 25 percent of water consumption. However, it could have a much larger impact on the Basin with the onset of climate change.[1] Most climate change studies predict a warmer and wetter climate, on average, for the Basin. Such a climate could actually benefit agriculture by creating a longer growing season in northern areas, by creating opportunities for double-cropping in southern areas, and by increasing crop yields through CO_2 fertilization. However, this warmer and wetter average climate is also expected to be more variable than the current climate. That means more intensive precipitation episodes and longer dry periods during the summer (Mortsch et al., 2003). This climatic variability could create a real problem for Basin farmers who mostly rely on relatively dependable spring and summer rainfalls. Increased use of irrigation would be one means of coping with increased climatic variability, providing farmers with dependable water in the face of erratic rainfalls. Multiple reports on the potential impacts of climate change on the Basin have noted this fact and predicted increased use of irrigation as one likely outcome from climate change (Mortsch et al., 2003; Penney, 2012).

Irrigation has the potential to preserve and enhance Great Lakes agriculture in the context of climate change, but it also has the potential to deplete local water sources and create intractable conflicts among competing water uses. Irrigation can be a voracious consumer of water, involving large takings with little return flow (Postel, 1999). In places throughout the western United States and Canada where large-scale irrigation has existed for decades, conflicts between irrigation and other human water uses are endemic, as are conflicts between irrigators and environmentalists with clashing worldviews on the nature and purpose of water resources (Worster, 1985; Reisner, 1993; Heinmiller, 2016). All these conflicts require careful balancing and management on the part of government regulators who are usually caught in the middle and pressured by all sides. Thus, there is a need to document emerging trends in Great Lakes irrigation and to investigate the potential for irrigation-based water conflicts as climate change takes hold.

Methods

To identify potential irrigation-related challenges of the future, this chapter examines the recent past. Based on data from the Great Lakes Commission's *Annual Report of the Great Lakes Regional Water Use Database*, it documents where the Basin's irrigation hotspots are located and identifies trends in irrigation water use. It does so by comparing water use data from 1998 and 2012, two years with similar rainfall levels during their summer growing seasons: 288 millimeters (11.3 inches) and 300 millimeters (11.8 inches), respectively. Both seasons are drier than normal for the Basin, so they are seasons in which Great Lakes farmers were more likely to rely on irrigation to supplement rainfall. A comparison of these years provides a good idea of whether Great Lakes farmers rely more on irrigation now than in the recent past. After establishing these trends, the chapter presents a case study of the most significant irrigation-based conflict in the Great Lakes Basin to date: the battle over the Mud Creek Irrigation District in Michigan.

Findings

Trends in irrigation water use

The 2012 data[2] in Table 3.1 show that four jurisdictions account for about 90 percent of the irrigation in the Great Lakes Basin: Michigan, Ontario, Wisconsin, and Indiana. The Basin's irrigation hotspots are also found in these four jurisdictions, as shown in Figure 3.1, and each is worthy of description.

Michigan is, by far, the largest irrigator in the Basin. In most years, it withdraws more irrigation water than all the other states and provinces combined. Michigan's reliance on irrigation is explained by a combination of factors. According to

Table 3.1 Irrigation water withdrawals in the Great Lakes Basin: 1998 vs. 2012

Jurisdiction	*1998 (million gallons/day)*	*2012 (million gallons/day)*	*Percentage Change*
Illinois	0	0.05	—
Indiana	25.64	67.79	164%
Michigan	273.12	416.28	52%
Minnesota	0.38	0.24	—
New York	3.6	37.95	954%
Ohio	14.79	30.99	110%
Ontario	68.46	73.22*	7%*
Pennsylvania	0.35	0.13	—
Quebec	9.22	3.02	-67%
Wisconsin	47.55	60.23	27%
Basin Irrigation	443.11	689.9	56%

* In 2011, Ontario changed the way it reports its irrigation withdrawals to the Great Lakes Commission, and the new reporting captures only a very small portion of irrigation in the province. The statistic for Ontario is taken from 2010 and probably underestimates the province's irrigation withdrawals in 2012.

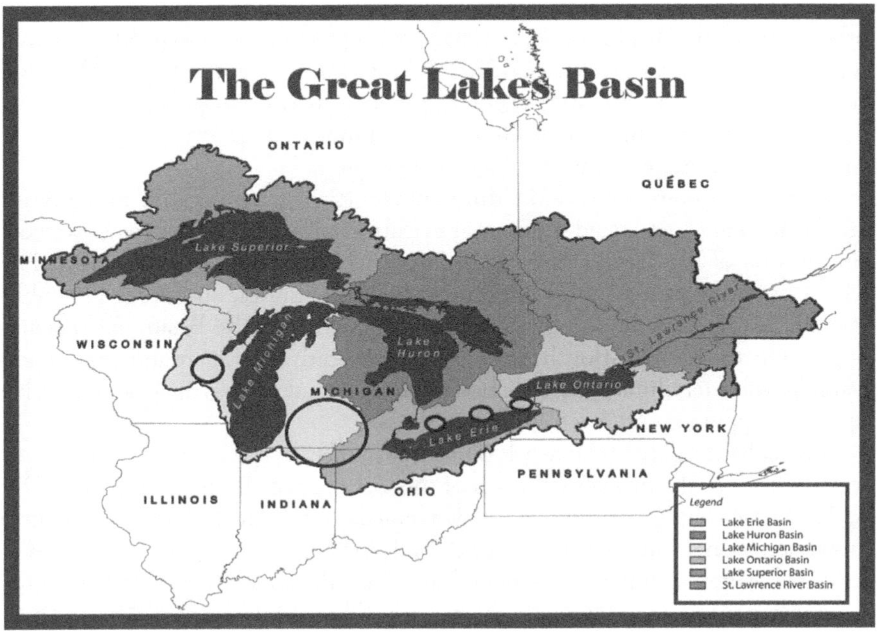

Figure 3.1 Irrigation hotspots in the Great Lakes Basin.

rainfall records, it is "the driest state east of the Mississippi River during the critical growing months of July and August" (Michigan Department of Agriculture and Rural Development, 2012: 13). Moreover, it is the only jurisdiction whose territory lies entirely within the Great Lakes Basin, so all its irrigation withdrawals come from Basin water. Irrigation is mainly concentrated in the southwestern part of the state, particularly around Kalamazoo, and is used primarily to grow corn, soybeans, vegetables, potatoes, and nursery crops (Michigan Department of Environmental Quality, 2006). There is also some irrigation in the Thumb area of the Lower Peninsula northeast of Saginaw, which is home to the Mud Creek Irrigation District, a project at the center of the significant irrigation-related water conflict that this chapter examines in detail

Ontario is the Basin's second-largest irrigator and, similarly to Michigan, draws almost all its irrigation water from the Great Lakes Basin because most of its arable land coincides with the Basin's hydrological boundaries. Irrigation is concentrated in the southern part of the province, particularly the Niagara Peninsula, the Norfolk Sand Plain, the Essex Region, and the southern shores of Lake Simcoe, where it is used to grow a wide variety of fruit and vegetable crops, and, to a lesser extent, field crops (de Loë et al., 2001; Statistics Canada, 2011). Most irrigators in Ontario are self-supplied – that is, they pump their own water from nearby ground or surface water sources – but some are supplied through public water systems. This is the case for irrigators in the Niagara-on-the-Lake area where, during drought conditions in the summer of 2016, irrigators from other parts of the Niagara region pressed local politicians to expand the public supply system (Edwards, 2016). Thus, there is potential for considerable irrigation growth in Ontario, and recent experience in the Niagara region may be illustrative of the political pressure that could mount for irrigation expansion as the effects of climate change are increasingly felt.

Wisconsin and Indiana are the third- and fourth-largest irrigators in the Basin, and both have experienced substantial irrigation growth in recent years. Wisconsin has one of the Basin's irrigation hotspots, located southwest of Green Bay, where irrigation is used to grow corn, soybeans, potatoes, and other vegetables (Smail, 2015). Indiana occupies only a small portion of the Basin, but this area is adjacent to the irrigation hotspot in southwestern Michigan and experiences a similar summer climate, so Indiana farmers have taken to irrigation, and it has become much more prevalent in recent years.

Judging by the data in Table 3.1, there seems to be an upward trend in irrigation water withdrawals throughout much of the Great Lakes Basin. Comparing 1998 to 2012 – two years with similar below-average rainfall levels during the summer growing season – total irrigation withdrawals increased by 56 percent. Moreover, irrigation water withdrawals increased in all jurisdictions where irrigation is not insignificant: Indiana, Michigan, New York, Ohio, Ontario, and Wisconsin all registered substantial increases in irrigation withdrawals during this time. New York had by far the largest percentage increase in withdrawals, mostly because it started from a small baseline, so any substantial increases would provide a manifold percentage increase. Quebec was the only jurisdiction to report a decline in irrigation withdrawals for reasons that are not entirely clear. There is no question that the overall regional trend is toward increased irrigation withdrawals, and mostly large increases at that.

This suggests that governments and farmers in the Great Lakes Basin may be turning to irrigation to adapt agricultural practices to the changing climate. According to the National Oceanographic and Atmospheric Administration (NOAA), the 15 warmest years on record have all taken place since 1998 (NOAA National Centers for Environmental Information, 2016). Thus, the effects of climate change have been felt, and it is probably no coincidence that irrigation water withdrawals also increased markedly during this period. Thus far, irrigation-based water conflicts have been relatively rare in the Basin, but there is one example from recent history that may provide a window on what is to come.

The Mud Creek Irrigation District conflict

Completed in 1996, the Mud Creek Irrigation District was Michigan's first – and, so far, only – irrigation district. The district is in an area vulnerable to extended dry periods during the growing season. The Thumb has a long history of irrigation, with many farmers self-supplying from local groundwater sources and small creeks. However, in particularly dry seasons the demand for water outstrips supply, and farmers recognized that they could expand their irrigated acreage and improve their crop yields if they could gain access to more water. Such a water supply was right on their doorstep in the form of Lake Huron, which surrounds the Thumb on three sides (Annin, 2006).

In the early 1980s, a small group of farmers approached the Michigan Department of Natural Resources with a plan to form an irrigation district and supply it by pumping water out of Lake Huron and piping it to their lands (Annin, 2006). Over the next decade, this plan evolved into a scheme in which Mud Creek, a small creek flowing into Lake Huron, would be excavated and deepened to reverse its flow, bringing 32.5 million liters (8.6 million gallons) of water daily from Lake Huron to the new irrigation district (Annin, 2006). This project was completed in 1996 and became the subject of considerable political conflict, both within Michigan and across the Basin.

For the most part, irrigation and irrigation water withdrawals fall under the jurisdiction of the state and provincial governments in the Basin. Just about all of the irrigation water in the Basin is supplied from groundwater or surface water tributaries of the Great Lakes, and it is the state and provincial governments that regulate water withdrawals from these sources. If an irrigator were to withdraw water from a boundary water – a water body through which the international boundary runs – and this withdrawal is large enough to noticeably affect water levels on the other side of the border, then the federal governments and the International Joint Commission would get involved. However, this is largely theoretical because it is difficult to contemplate an irrigation project large enough to withdraw that much water (Heinmiller, Forthcoming). For example, the Mud Creek irrigation withdrawal was estimated to drop the level of Lake Huron by 0.0000455 feet (0.013 millimeters), which made it a state rather than a federal concern (Annin, 2006). The states and provinces also have jurisdiction over the creation of irrigation districts, though Michigan is one of the few Great Lakes states with an irrigation district law. This law dates to the late 1960s and was created during a particularly bad drought to facilitate more irrigation. It is this law that a small group of farmers in the Thumb used to form the Mud Creek Irrigation District (Annin, 2006).

Recognizing their hydrological interdependence, the Great Lakes states and provinces have also negotiated a series of intergovernmental agreements to deal with the regulation of particularly large consumptive water uses, such as irrigation. At the time the Mud Creek proposal was coming to fruition, the operative intergovernmental agreement was the Great Lakes Charter, signed in 1985. The Charter created an intergovernmental procedure for review of a specific class of water withdrawals regarded by the signatory governments as threatening. All proposed water withdrawals in excess of 19 million liters (5 million gallons) per day and intended for a consumptive use – a use that returns little or no water to its source – such as irrigation would be subjected to a Prior Notice and Consultation Process involving all the Basin governments. Initiation of the process was made the responsibility of the permitting jurisdiction, and all Basin governments were to have a chance to comment and object to a prospective water removal, although the final decision remained with the permitting government. Other commitments were made to make changes in the regulation of smaller water-takings, but the Prior Notice and Consultation Process was the centerpiece of the Charter and played a central role in the Mud Creek irrigation controversy (Heinmiller, 2007).

Though Michigan was responsible for regulating the creation of the district and approving the increased water withdrawal, federal money made the project possible. The farmers behind the project found an important ally in their local U.S. representative, Bob Traxler, who also was an influential member of the House Appropriations and Agriculture committees. Traxler repeatedly directed federal money to the project to fund feasibility studies, to support completion of the necessary state applications, and, ultimately, to build the project. Altogether, around US$2 million of federal money was spent on the Mud Creek project and, without this funding, the project probably would have died at an early stage. Just over a dozen farmers were involved in the project, and it is unlikely that they could have shouldered its financial burden on their own (Annin, 2006).

Serious conflict began after Michigan notified the other Great Lakes jurisdictions of the Mud Creek water withdrawal proposal in November 1992, as it was obligated to do under the Charter's Prior Notice and Consultation process. Until this point, the Mud Creek process had mainly involved a back-and-forth between the farmers proposing the district and the state Department of Natural Resources, addressing feasibility issues. However, once the other governments were notified of the proposal, it drew criticism and opposition from various quarters.

Environmental groups opposed the project because they regarded it as an unnecessary and unsustainable use of Great Lakes water. For example, Great Lakes United, the largest Great Lakes environmentalist advocacy group at the time, opposed the project during the consultation stage. It argued that the water being withdrawn would be an environmental waste because the crops being grown were, on a national scale, already in surplus. They also expressed fear that the project would set a bad precedent for the Basin, replicating irrigation-related environmental problems already evident in the American West (Annin, 2006).

Many state and provincial governments were just as critical of the Mud Creek project. After receiving the initial notification, six premiers and governors demanded a consultation meeting on the project. One took place in April 1993, where they expressed their objections. The proposal was particularly galling for the other governments because it came less than a year after Michigan had vetoed an inter-Basin diversion proposal to supply water to the town of Lowell, Indiana, and the Lowell proposal had involved only a fraction of the water in the Mud Creek proposal.[3] Less than two weeks after the consultation meeting, Michigan Governor John Engler notified the other governments that, despite their objections, he would recommend approval of the project to the Michigan Natural Resources Commission,[4] which had the final say on the matter. This provoked another angry round of objections from the other governments but, since the Charter required Michigan only to consult the other governments, the objectors had no recourse to block Mud Creek's approval (Heinmiller, 2007). One month later, in June 1993, the Michigan commission approved Mud Creek's application, and the district was created.

The Mud Creek saga illustrates how irrigation-related conflicts can develop in the Great Lakes Basin. Irrigation usually involves large volumes of water, most of which is consumed, giving them a zero-sum dimension. As a result, water users sharing a resource with irrigators, such as the other Great Lakes states and provinces, often object to irrigation expansion because they feel they are losing out. Similarly, environmentalists often object to irrigation expansion because it leaves less water to support ecosystem functions and may contribute to pollution. As irrigation water withdrawals increase in the Basin, conflicts of these sorts are likely to become more common, both within jurisdictions and across the jurisdictions sharing the Basin.

Managing irrigation-based conflicts in the future

For most states and provinces, the Mud Creek conflict highlighted some glaring weaknesses in the Great Lakes Charter and prompted new efforts by the governors and premiers to create stronger intergovernmental water governance institutions. In June 2001, they negotiated a charter amendment known as the Charter Annex that committed them to work toward a binding Basin-wide agreement to preserve and protect the Basin's water for the use and benefit of its residents. This prompted another round of negotiations, eventually resulting in the *Great Lakes-St. Lawrence River Basin Sustainable Water Resources Agreement* (the agreement) and its companion *Great Lakes-St. Lawrence River Basin Water Resources Compact* (the compact) in December 2005 (Council of Great Lakes Governors). The two documents are substantively similar, but the compact uses the U.S. Constitution's interstate compact mechanism to create an institution in which courts can enforce state compliance.[5] Lacking a comparable compact mechanism under the Canadian Constitution and unable to join the U.S. compact under the U.S. Constitution, the agreement commits Ontario and Quebec to the same water conservation measures as their U.S. counterparts but in a political handshake agreement (Hall, 2006). For irrigation, both documents provide a much different institutional context than prevailed during the Mud Creek conflict.

A central feature of the agreement/compact is the Great Lakes governments' commitment to a "decision-making standard" for regulating water withdrawals within the Basin. The agreement and compact require signatory governments to establish volumetric threshold levels for regulation of withdrawals and to apply the decision-making standard to all new or increased withdrawals above the threshold. The precise level of the regulatory threshold is undefined in the agreement/compact, but each government must define it so overall water uses are reasonable and do not result in significant adverse impacts, thus achieving the documents' objectives. Should any government fail to establish a regulatory threshold within five years, a threshold of 379,000 liters (100,000 gallons) per day automatically takes effect, providing some indication of what the framers had in mind as a reasonable threshold (Hall, 2006).[6] The intent was to cast a wider regulatory net for water withdrawals across the Basin, one that would catch many new or increased water-takings for irrigation, subjecting them to the provisions of the decision-making standard.

In scope, the decision-making standard applies to all new or increased water withdrawals from any water source in the Basin, including the lakes themselves, lake tributaries, and even groundwater. The inclusion of all the Basin's surface and groundwater sources under a single regulatory umbrella is unprecedented in the Basin and, according to Hall, "is a long overdue advancement in water law" (Hall, 2006: 435). It also means that all prospective sources of future irrigation water are covered by the agreement/compact, again reinforcing the relevance of the decision-making standard in the regulation of future irrigation growth. That standard does not apply to existing water withdrawals – their regulation is left to the discretion of the respective states/provinces – but the agreement/compact does commit the states/provinces to implement water conservation programs for all users, including existing users (Hall, 2006).

Substantively, the decision-making standard establishes five principles that states and provinces must follow in evaluating proposed new or increased withdrawals from Basin waters. The first four principles are based on concerns with source water protection and past experiences with transboundary regulation of water-takings in the Basin, while the fifth principle has its roots in common law riparian rights and the reasonable use doctrine (Hall, 2006). Stated concisely, the standard's five principles are:

1 All withdrawals will be returned to the source watershed, less an allowance for consumptive use.
2 All withdrawals, individually or cumulative, will not have a significant adverse impact on source watersheds.
3 All withdrawals will use environmentally sound and economically feasible water conservation measures.
4 All withdrawals will comply with applicable municipal, state, and federal laws, as well as relevant interstate and international agreements.
5 All withdrawals will make reasonable use of the water based on defined criteria such as: the efficiency of the proposed use; the efficiency of existing uses; the balance among economic development, social development, and environmental protection; the capability of a source to supply water; the adverse impacts on other uses; and any potential restoration of degraded sources (Great Lakes – St. Lawrence River Basin Sustainable Waters Agreement, 2005).

Even a cursory reading of these principles suggests that they set the bar pretty high in terms of standards for new or increased water withdrawals. These principles place the burden on prospective irrigators to show that their withdrawals will not have significant adverse impacts on water sources, that they use sound conservation practices, that they adhere to relevant laws, and that they are reasonable in all specified senses of the term. This does not make irrigation expansion impossible by any means. Rather, it lays out a precautionary approach to irrigation expansion that guards against environmental degradation, wasteful water use practices, and impairments of existing uses. This significantly changes the political playing field from what it was during the Mud Creek controversy, providing some empowerment to water users and environmentalists who may feel threatened by irrigation expansion.

Discussion and conclusions

Over the past 20 years, the climate of the Great Lakes Basin has shown signs of warming, with longer and hotter dry spells in the summer becoming more common. Water withdrawal data indicate that some farmers in the Basin have turned to irrigation to adapt to the hotter climate, with irrigation water withdrawals increasing by 56 percent in 2012 as compared to 1998. As more farmers turn to irrigation, the potential for irrigation-related conflicts also increases. Irrigation consumes large volumes of water and, as the Mud Creek case shows, this consumption can result in conflicts with other users seeking to protect their own water access and with environmentalists who want water left in the natural environment to protect ecosystem functions. In the Mud Creek case, the intergovernmental institutions regulating irrigation were skewed in the irrigators' favor, but changes adopted in the 2005 compact/agreement seem to have levelled the political playing field. Any future irrigation-related conflicts will now play out in a context in which irrigators, and the governments supporting them, must justify their withdrawals as necessary, efficient, and sustainable. How this will look in practice is uncertain, as a major irrigation conflict has yet to test these institutions, but it seems a matter of time before such a test comes along.

Notes

1 The water use data are based on an analysis of data in the Great Lakes Commission's *Annual Report of the Great Lakes Regional Water Use Database* for 2006 to 2014, inclusive.

2 Climatic conditions in 2012 and 1998 match pretty closely, setting up the comparison of irrigation water use between these two years. With more current data, climatic conditions would not match as well and the comparison would be much less effective.

3 Michigan had authority to veto the Lowell proposal because of provisions in the federal Water Resources Development Act (WRDA) that allow all Great Lakes states to veto any diversions of Great Lakes water outside of the Basin. Since Lowell is just outside the Basin's boundaries, WRDA applied; but since the Mud Creek project was well inside the boundaries, WRDA did not apply. See Heinmiller (2007) for more details.

4 Michigan's Natural Resources Commission is an arms-length decision-making body whose members are appointed by the governor.

5 For the compact to take effect and become legally binding, it had to be passed into law in all eight U.S. states and the U.S. Congress. This happened in surprisingly rapid fashion starting with Minnesota in February 2007 and ending with Congress in October 2008.

6 The signatory governments are also required to create water resource inventories that document all available water resources and all withdrawals of 379,000 liters (100,000 gallons) per day or greater. So again, the 379,000 liters (100,000 gallons) per day threshold seems to be whatr the agreement/compact's framers had in mind as an appropriate regulatory threshold.

References

Annin, P. (2006). *The Great Lakes Water Wars*. Washington, DC: Island Press.

de Loë, R., Kreutzweiser, R., and Ivey, J. (2001). Agricultural water use in Ontario. *Canadian Water Resources Journal*, 26(1): 17–42.

Edwards, L. (2016, July 28). Politicians, farm leaders talk irrigation needs for Niagara. *Niagara This Week*, p. 73.

Great Lakes – St. Lawrence River Basin Sustainable Waters Agreement. (2005). Available at www.glslregionalbody.org/GLSLRBAgreements.aspx

Hall, N. D. (2006). Toward a new horizontal federalism: Interstate water management in the Great Lakes region. *University of Colorado Law Review*, 77: 405–456.

Heinmiller, B. T. (2007). Do intergovernmental institutions matter? The case of water diversion regulation in the Great Lakes Basin. *Governance*, 20(4): 655–674.

Heinmiller, B. T. (2016). *Water Policy Reform in Southern Alberta: An Advocacy Coalition Approach*. Toronto: University of Toronto Press.

Heinmiller, B. T. (Forthcoming). Institutions and transboundary governance capacity in the Great Lakes Basin: The case of irrigation water-takings. *International Journal of Water Governance*.

Kling, G. W., Kayhoe, K., Johnson, L. B., Magnuson, J. J., Polasky, S., Robinson, S. K. et al. (2003). *Confronting Climate Change in the Great Lakes Region: Impacts on Our Communities and Ecosystems*. Cambridge, MA: The Union of Concerned Scientists and the Ecology Society of America.

Michigan Department of Agriculture and Rural Development. (2012). *Generally Accepted Agricultural and Management Practices for Irrigation Water Use*. Available at michigan.gov/documents/MDA_Irrigation_GAAMP_129710_7.pdff

Michigan Department of Environmental Quality. (2006). *Water Withdrawals for Agricultural Irrigation in Michigan 2006*.

Mortsch, L., Alden, M., and Scherago, J. (2003). *Climate Change and Water Quality in the Great Lakes Region: Risks, Opportunities and Responses*. A Report Prepared for the Great Lakes Water Quality Board of the IJC.

NOAA National Centers for Environmental Information. (2016). *State of the Climate: Global Analysis for Annual 2015*. Available at www.ncdc.noaa.gov/sotc/global/201513

Penney, J. (2012). *Adapting to Climate Change: Challenges for Niagara*. Available at https://glslcities.org/library/adapting-to-climate-change-challenges-for-niagara/

Postel, S. (1999). *Pillar of Sand – Can the Irrigation Miracle Last?* New York: W.W. Norton & Company.

Reisner, M. (1993). *Cadillac Desert – The American West and Its Disappearing Water*. New York: Penguin.

Smail, B. (2015). *Wisconsin Water Use – Agricultural Irrigation Rates*. Available at state.awra.org/wisconsin/2015meeting/Session3B1Smail.pdf

Statistics Canada. (2011). *Agricultural Water Use in Canada 2010*. Catalogue no. 16-402-X, Ministry of Industry. Ottawa: Government of Canada.

Worster, D. (1985). *Rivers of Empire – Water, Aridity and the Growth of the American West*. New York: Oxford University Press.

4 Artificial reefs and reef restoration in the Laurentian Great Lakes

Edward F. Roseman, Matthew McLean, Jeremy J. Pritt, Jason L. Fischer, and *Gregory Kennedy*

Introduction

Habitat loss and degradation are globally recognized contributors to fisheries declines (e.g., Jelks et al., 2008; Munday, 2004; Powels et al., 2000), and restoration of physical habitat is often used to mitigate those losses (Palmer et al., 2005). Artificial reefs have been constructed in freshwater and marine systems to (1) attract fish (Creque et al., 2006; Kelch, 2012), (2) provide spawning, nursery, and adult habitats, and (3) increase fish abundance (Brickhill et al., 2005; Dumont et al., 2011; Kuhl, 1992; Pickering and Whitmarsh, 1997). Artificial reefs have been proven to attract fish and increase catch rates in recreational fisheries (Bombace et al., 1994; Brickhill et al., 2005; Kelch, 2012), but the ability of reefs to increase fish diversity and numerical abundance is not well documented in freshwater and marine systems (Baine, 2001; Pickering and Whitmarsh, 1997).

Researchers reviewed artificial reef projects in marine and freshwater systems and determined that assessment of reefs seldom detected an increase in fish diversity or population abundance. Critiques arose about the results that included flawed study design (Brickhill et al., 2005), lack of clear objectives for reef-building (Baine, 2001; Claudet and Pelletier, 2004), and inadequate monitoring (Baine, 2001). Few studies concluded that artificial reefs can perform as well or better than natural reefs in terms of increasing fish abundance and richness at the reef site (Carr and Hixon, 1997; Koeck et al., 2014). However, some artificial reefs were associated with fish communities differing in composition from those in natural habitats (Ambrose and Swarbrick, 1989; Flopp et al., 2013). It remains unclear whether artificial reefs can function as well as natural reefs from an ecological perspective (Perkol-Finkel et al., 2006).

Despite careful planning, artificial reefs can be damaged or degraded, limiting their usefulness as spawning/rearing habitat over time and underscoring the need for long-term monitoring (McLean et al., 2014). Damage and degradation can be caused by navigation and channelization, sedimentation, intrusion of groundwater, and changes in water levels (Rutherford et al., 2004). Other potential threats to reef quality and function include redistribution of reef materials by wave action, fouling by invasive species (e.g., *Dreissenidae*), and algae accumulation (Marsden and Chotkowski, 2001; Johnson et al., 2006). These factors often limit the effectiveness of reefs. Many stressors do not pose an immediate threat but develop over time, emphasizing the need for long-term assessment (Gannon,

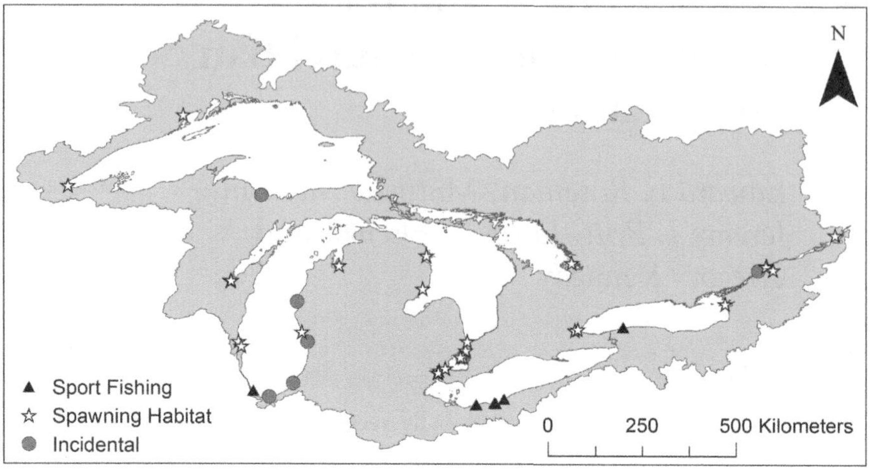

Figure 4.1 Locations of the artificial reef sites discovered in the Great Lakes and Great Lakes connecting channels and tributaries.

Map: Google Maps. Figure updated and adapted from McLean et al. (2014).

1990; McLean et al., 2014). Similarly, strategies and techniques for reef maintenance are rarely considered in artificial reef project planning, but could provide longevity and improved performance of reefs after the initial investment.

Statement of the problem

In the Laurentian Great Lakes, artificial reefs were created to provide spawning and nursery habitat to lithophilic spawning fishes (Fitzsimons, 1996; Gannon, 1990; NOAA, 2007). Artificial reefs have been used as a fishery management tool for more than 40 years in the Laurentian Great Lakes and have often been championed as a successful means for improving recreational fisheries (Kelch, 2012). McLean et al. (2014) conducted a literature review of artificial reef projects in the Great Lakes to guide future habitat restoration and monitoring efforts. They documented 31 constructed reefs and 7 incidental (manmade without intent to create habitat, such as coal cinders dumped from coal-burning ships) artificial reefs (Figure 4.1; Table 4.1). Objectives for construction included reducing impacts of currents and waves, providing safe harbors, improving sport-fishing opportunities, and enhancing/restoring fish spawning habitats (McLean et al., 2014). This chapter updates their inventory, adding newly discovered and created projects, and further discussing the need for monitoring and maintenance of artificial reefs.

Methods

We updated McLean et al.'s (2014) literature review by adding newly discovered and constructed reefs to the inventory. For all relevant Great Lakes reef studies,

Table 4.1 Artificial reefs located in the Laurentian Great Lakes, tributaries to the St. Clair and Detroit River System (SCDRS), and St. Lawrence River (SL) connecting channels

Type	*Reef Site*	*Year Created*	*Material*	*Purpose*	*Biological Assessment*	*Physical Assessment*
Connecting Channels	Belle Isle (SCDRS)	2004	limestone, cobble, coal cinders	fish spawning	4 yrs.	none
	Fighting Island (SCDRS)	2008	limestone, fieldstone	fish spawning	ongoing	none
	Northeast Belle Isle (SCDRS)	2013	limestone, fieldstone	fish spawning	3 yrs. but ongoing	1 yr.
	Algonac (SCDRS)	Late 1800s	coal cinders	incidental reef	2 yrs. but ongoing	2 yrs.
	Middle Channel (SCDRS)	2012	limestone, fieldstone	fish spawning	ongoing	none
	Port Huron (SCDRS)	2012	limestone	fish spawning	none	2 yrs.
	Harts Light (SCDRS)	2015	limestone	fish spawning	2 yrs.	2 yrs.
	Pointe Au Chenes (SCDRS)	2014	limestone	fish spawning	2 yrs.	2 yrs.
	Grassy Island (SCDRS)	2015	limestone	fish spawning	2 yrs.	2 yrs.
	Fort Wayne (SCDRS)	2015	limestone	fish spawning	2 yrs.	2 yrs.
	Sythe Energies (SL)	1993	limestone	incidental reef	3 yrs.	diver visual
	Iroquois S. (SL)	2007	gravel and boulders	fish spawning	2 yrs.	1 yr.
	Iroquois N. (SL)	2007	gravel and boulders	fish spawning	2 yrs.	1 yr.
Lake Erie	Polish Fisherman's Club Reef, Lorain, OH	1986–1989	clean rock, rubble, concrete, brick scrap	sport fishing	2 yrs.	none
	Mountain Reef, Lorain, OH	1986–1989	clean rock, rubble, concrete, brick scrap	sport fishing	2 yrs.	none
	Cuyahoga County Commissioners Reef, Lakewood, OH	1986–1989	clean rock, rubble, concrete, brick scrap	sport fishing	2 yrs.	none

(*Continued*)

Table 4.1 (Continued)

Type	*Reef Site*	*Year Created*	*Material*	*Purpose*	*Biological Assessment*	*Physical Assessment*
	Edgewater Artificial Reef "A" (North)	1998	old stadium rubble	sport fishing	none	none
	Edgewater Artificial Reef "B" (South)	1998	old stadium rubble	sport fishing	none	none
	Euclid Artificial Reef "C"	1998	old stadium rubble	sport fishing	none	none
Lake Huron	Tawas	1987	limestone	fish spawning	7 yrs.	none
	Thunder Bay, Cement Plant Array	2010–2011	limestone	fish spawning	6 yrs. planned	—
	Thunder Bay CKD Pile Array	2010–2011	limestone	fish spawning	6 yrs.	—
	Severn Sound		rock, gravel	fish spawning	none	none
Lake Michigan	Cook Nuclear Plant, Bridgman, MI		riprap	incidental reef	1 yr.	none
	J.H Campbell Power Plant		riprap	incidental reef	2 yrs.	none
	Ludington Power Facility, MI			incidental reef	2 yrs.	none
	South Shore Reef Park, Milwaukee, WI		field stone/riprap	fish spawning	2 yrs.	3 yrs.
	Hamilton Reef, Muskegon, MI		cement/rubble	fish spawning	2 yrs.	2 yrs.
	Port of Indiana Breakwater	1992–1996	large and small stones (limestone)	incidental reef	5 yrs.	5 yrs.
	Chicago Waterfront	1999	granite rubble	sport fishing	4 yrs.	none
	Oak Creek Reef	2007	limestone	fish spawning	5 yrs.	none
	Grand Traverse Bay	2015	field stone	fish spawning	2 yrs.	diver visual
Lake Ontario	Western Lake Ontario	1981–1982	red shale, red siltstone	sport fishing	2 yrs.	none

	Stony Island	1989	limestone rock	fish spawning	2 yrs.	none
	Hamilton Harbor		rock, gravel	fish spawning	none	none
	Hamilton Harbor Shoal	2010	rock and cobble	fish spawning	5 yrs.	none
Lake Superior	Presque Isle Harbor	1974–1977	granite and limestone	incidental reef	4 yrs.	4 yrs.
	Fitgers Reef	1990	boulder and cobble	fish spawning	3 yrs.	none
Tributaries	Des Prairies River (St. Lawrence River tributary)	1996	canal excavation material	fish spawning	7 yrs.	4 yrs.
	Grasse River (St. Lawrence River tributary)	1996	limestone shot rock	fish spawning	none	none
	Current River (L. Superior tributary)	1991	igneous rock	fish spawning	3 yrs.	none
	Brown County Fairgrounds (Fox River, Lake Michigan tributary)	1989	dolomite limestone	fish spawning	6 yrs.	none
	Voyageur Park (Fox River, Lake Michigan tributary)	1990	dolomite limestone	fish spawning	5 yrs.	none
	Fox Point Marina (Fox River, Lake Michigan tributary)	1992	dolomite limestone	fish spawning	3 yrs.	none

Adapted from McLean et al. (2014).

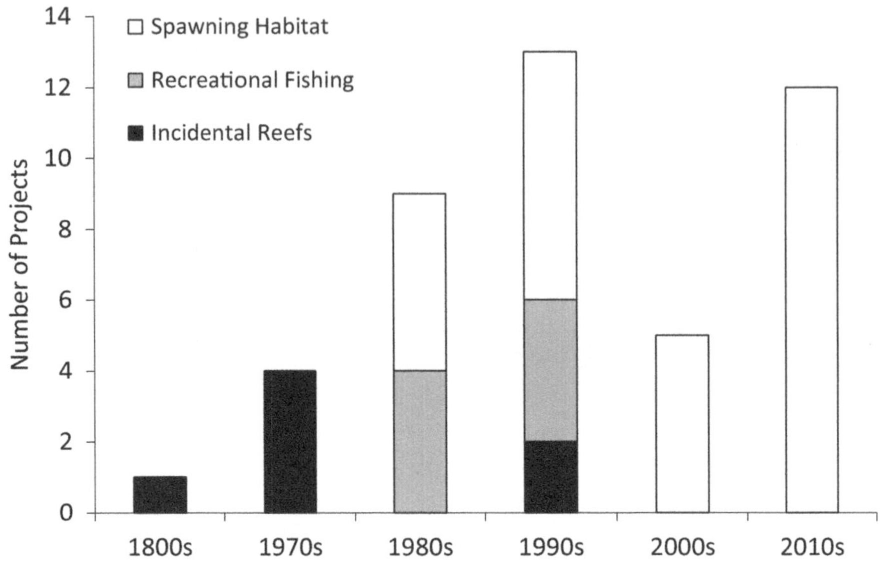

Figure 4.2 Number, types, and timing of artificial reef and structure creation in the Great Lakes.

Figure updated and adapted from McLean et al. (2014).

we obtained reef site name, location, year of construction, objectives for construction, materials used, monitoring scope, and timing/duration of monitoring (Table 4.1).

Findings

Since 2000, the focus of artificial spawning habitat construction has shifted from protective physical habitat structures to restoring and increasing functional spawning habitat for the enhancement of Great Lakes fish populations (Dumont et al., 2011; Fitzsimons, 1996; Manny et al., 2010; Roseman et al., 2011) (Figure 4.2). These projects emphasized creation of spawning habitat for imperiled or important species, such as lake sturgeon (Dumont et al., 2011; Roseman et al., 2011), lake trout (Fitzsimons, 1996; Marsden et al., 2016), and walleye (Geiling et al., 1996), or improving diversity of benthic-spawning fishes (Manny et al., 2010). Older reefs were built in areas to provide access to anglers (Kelch, 1997, 2012; Creque et al., 2006), but recent reefs have been built in known fish spawning areas, such as channels connecting the Great Lakes and tributary rivers.

Biological monitoring of reef projects

Based on findings reported in McLean et al. (2014), biological monitoring of artificial reefs in the Great Lakes typically lasts 2 to 7 years after construction (mean of 2.9 years across completed studies; see Table 4.1). The focus of biological monitoring has been either the abundance/diversity of adult fish (Fitzsimons, 1996; Kelch et al., 1999; Creque et al., 2006; Dumont et al., 2011; Environment Illimité Inc., 2009; Geiling et al., 1996; Marsden et al., 2016) or the early life history stages to gauge the reproductive success of spawning fish (Bouckaert et al., 2014; Manny et al., 2010; Marsden and Chotkowski, 2001; Nichols et al., 2003; Roseman et al., 2011; Schreiner et al., 1995). Many monitoring studies lack biological or physical assessment before reef construction to compare to post-construction conditions (Figure 4.3; McLean et al., 2014). Surveys of fisheries biologists (Geiling et al., 1996) and anglers (Creque et al., 2006; Kelch, 2012) have also been used to determine fish use of artificial reefs. Marranca et al. (2015) conducted a novel approach to examine the genetic effects of lake sturgeon spawning on artificial reefs in the St. Clair-Detroit River System (SCDRS) using estimates of effective number of breeders per reef and measures of genetic diversity of reef-origin eggs and larvae.

Physical monitoring of artificial reef projects

Most monitoring protocols for artificial reefs in the Great Lakes focused on the biological attributes of the reef, with physical aspects secondary or not measured at all (McLean et al., 2014; see Table 4.1). Scuba divers, underwater video, and remotely operated vehicles (ROVs) were most frequently used for nonstandardized, qualitative observations. Methods for quantifying physical habitat attributes

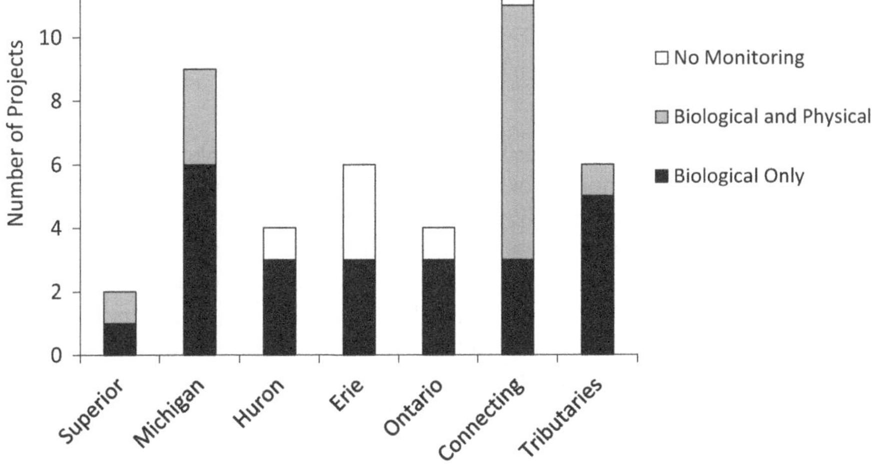

Figure 4.3 Artificial reef monitoring conducted in the Great Lakes.

Figure updated and adapted from McLean et al. (2014).

included using scuba divers to map reefs (Binkowski, 1985); measure flow velocity, depth, and substrate (Dumont et al., 2011); zebra mussel accumulation (Marsden and Chotkowski, 2001); and the amount of algal colonization on reefs (Rutecki et al., 1985). Sidescan sonar was used to evaluate the condition and movement of reefs (Collier and Brown, 2004; Edsall et al., 1989, 1992) and produce maps of reefs and adjacent areas (Collier and Brown, 2004; Kelch, 2012).

Combinations of physical habitat measurement techniques provide robust assessments of reef performance and enhanced ability to model reef design under varying hydrologic regimes. Sidescan sonar, acoustic Doppler current profilers (ADCP), flow modeling, underwater video, PONAR grab sediment sampling, particle size analysis, and scuba diver observations are being used to assess proposed reef sites and post-construction performance in the SCDRS (Fischer et al., 2015; Kinzel et al., 2016; Manny, 2006).

ADCP and flow modeling are relatively new technologies used in the development and monitoring of artificial reefs in large river systems where current plays a key part in the spawning activity of target species. In the SCDRS, flow models have been used to identify areas with water velocities used by spawning lake sturgeon and areas where sediment transport could compromise an artificial reef (Bennion and Manny, 2014; Kinzel et al., 2016). Direct measurements of water velocities were then obtained with an ADCP, which continuously measures water velocity as the instrument is moved along a transect. A mapping software program routine (rivSurveyR, github.com/jasonfischer/rivSurveyR) was used to map fine-scale flow patterns within an area of interest and make a final site selection (Fischer et al., 2015; Manny et al., 2015). Although these assessments have been directed at the pre-assessment phase of reef construction, they can provide information on changes in flow patterns and sediment movement, allowing managers to set goals for reef longevity. Repeated surveys could provide insight on how flow patterns influence reef maturation and how bathymetric measurements from ADCPs could be used to track physical changes in reef dimensions through time.

Degradation of artificial reefs: a need for monitoring and maintenance

Planning and construction of artificial reefs in the Great Lakes involve a large initial investment with hoped-for long-term benefits. McLean et al. (2014) found that some monitoring is typically conducted for most reefs, but it is usually limited in temporal scope and not multidisciplinary, thereby not providing information on performance and limiting the scientific ability to learn from an adaptive management approach (Manny et al., 2015). Few efforts to maintain the physical integrity of reefs have been attempted in the Great Lakes, likely due to lack of objectives related to long-term performance, restricted funding, and lack of established assessment and maintenance techniques.

Several approaches for reef maintenance have been attempted and assessed outside of the Great Lakes. Sediment removal techniques are used to clean substrates when infilling occurs. Flushing with high flows is used in regulated rivers, as well as naturally, when a spate occurs that greatly increases discharge and moves sediment downstream (Nelson et al., 1987). Agitation dredging is used to lift sediment from the target location and let current disperse or divert it from the target (Bass, 1985; Hotchkiss and Huang, 1995).

Addition of reef materials is commonly used to supplement and replenish habitats. For example, a common practice for restoration and maintenance of coral reefs includes transplanting corals to supplement communities and accelerate regeneration of damaged reefs (Harriott and Fisk, 1988; Lindahl, 2003; Spieler et al., 2001). In freshwater systems, gravel was added to augment salmon (*Oncho-rhyncus* spp.) spawning areas, and it successfully increased the number of redds and redd occupancy (Merz and Setka, 2004). Additions of gravel and rock to walleye spawning areas in rivers showed positive effects in egg deposition (Dustin and Jacobson, 2003; Geiling et al., 1996).

Physical degradation of artificial reefs can limit their long-term effectiveness as fish habitat. Accumulation of silt and colonization by *Cladophora* spp., and dreissenid mussels were documented to reduce fish use of reefs constructed in the Great Lakes (Johnson et al., 2006; Kevern et al., 1985; Marsden and Chotkowski, 2001; Peck, 1986). Thorough planning and pre-assessment evaluations allow managers to pinpoint construction/restoration efforts to areas with minimal amounts of sediment deposition or scouring flows that could displace reefs. However, given the dynamic nature of large lakes and rivers, infilling and movement are unavoidable. Therefore, projects that incorporate strategies for monitoring and maintenance will likely have the strongest long-term benefits.

Most artificial reef monitoring programs are short (i.e., < 5 years) and small scale (i.e., < 1 square kilometer), and may not match the life history of long-lived late-maturing species such as lake trout, lake sturgeon, walleye, and lake whitefish. Moreover, physical stressors such as sedimentation, degradation by waves and water currents, and water level fluctuations may occur over large spatial extents and longer time periods. Observance of fish population-level responses, such as increased abundance or changes in reef quality in response to stressors, would require monitoring a larger spatial extent and duration than most reef monitoring programs. Inadequate monitoring of artificial reefs is common (Smokorowski et al., 1998a) and problematic to understanding the results of artificial reef projects (Baine, 2001). Long-term monitoring of physical and biological attributes of constructed reefs is needed to measure biological effects and permanence of artificial reefs (Gannon, 1990; Hartig et al., 2009; McLean et al., 2014; NOAA, 2007). Long-term monitoring could improve fish habitat restoration techniques and better focus limited management resources through adaptive management (Palmer et al., 2005).

Reef placement, size, design, and the size and type of parent construction materials varied greatly among the 44 reef projects identified in our study (Table 4.1). Hence, our ability to discern the magnitude and significance of these confounding factors is limited. Future research is needed to determine the most effective materials and construction design that maximize the effectiveness of projects to meet specified objectives and match existing habitat conditions.

Although sampling methods depend on the habitat and fish species of interest, the wide variety of biological variables and techniques used to assess artificial reefs limits comparability of reef assessment studies and our understanding of artificial reefs as a fisheries management tool (Fitzsimons, 1996). Standardized monitoring protocols are needed to evaluate trends in reef performance through time (Bortone and Kimmel, 1991; Murray, 1994; NOAA, 2007; Sheng, 2000). Such standardized guidelines exist for reefs constructed outside of the Great Lakes and include the focal variables and duration of monitoring (e.g., Bortone and Kimmel, 1991; Lindberg and Seaman, 2011). Historically, physical habitat monitoring of artificial reefs in the Great Lakes has typically consisted of qualitative underwater observations, which may limit the ability to assess physical and biological changes over time. More quantitative underwater observations have been used to monitor physical attributes of artificial reefs outside the Great Lakes (e.g., Ambrose and Anderson, 1990; Wilding and Sayer, 2002). Standardized long-term monitoring of physical and biological variables on artificial reefs could improve the design and biological productivity of artificial reefs and allow fishery managers to better understand the benefits of reefs to fishery resources. We recommend that researchers focusing on Great Lakes reefs use quantitative methodology.

Conclusion

The use of artificial reefs as fisheries enhancement tools in the Great Lakes is widespread. Artificial reefs have been constructed in a wide range of habitats for many target species. Studies of artificial reefs in the Great Lakes using underwater cameras and sidescan sonar advanced the monitoring of biological and physical characteristics of artificial reefs (Foster and Kennedy, 1995; Manny, 2006). However, few of these methods have been incorporated into standardized, long-term monitoring programs.

We suggest the following practices for future research on the benefits of artificial reefs to fishery resources in the Great Lakes: (1) standardize biological and physical monitoring and methodology to allow for comparisons among reefs through time; (2) develop reef monitoring programs that better match the spatial and temporal extent of the biological responses and physical stressors; (3) monitor the physical, as well as biological changes in reefs; and (4) whenever practical, focus primarily on quantitative response variables, using qualitative assessment to fill temporal gaps in quantitative sampling. Following these recommendations should better quantify reef efficacy and performance, and improve our understanding of artificial fish spawning reefs.

References

Ambrose, R. F., and Swarbrick, S. L. (1989). Comparison of fish assemblages on artificial and natural reefs off the coast of southern California. *Bulletin of Marine Science*, *44*: 718–733.

Ambrose, R. F., and Anderson, T. W. (1990). Influence of an artificial reef on the surrounding infaunal community. *Marine Biology*, *107*: 41–52.

Baine, M. (2001). Artificial reefs: A review of their design, application, management and performance. *Ocean Coastal Management*, *44*: 241–259.

Bass, G. F. (1985). Archaeologists, sport divers, and treasure-hunters. *Journal of Field Archaeology*, *12*(2): 256–260.

Bennion, D. H., and Manny, B. A. (2014). A model to locate potential areas for lake sturgeon spawning habitat construction in the St. Clair – Detroit River System. *Journal of Great Lakes Research*, *40*: 43–51.

Binkowski, F. P. (1985). Utilization of artificial reefs in the inshore areas of Lake Michigan. In: D'Itri, F. M. (Ed.), *Artificial Reefs: Marine and Freshwater Applications*. Chelsea, MI: Lewis Publishers, 349–362.

Bombace, G., Fabi, G., Fiorentini, L., and Speranza, S. (1994). Analysis of the efficacy of artificial reefs located in different areas of the Adriatic Sea. *Bulletin of Marine Science*, 55: 559–580.

Bortone, S. A., and Kimmel, J. J. (1991). Environmental assessment and monitoring of artificial habitats. In: Seaman, W. and Sprague, L. M. (Eds.), *Artificial Habitats for Marine and Freshwater Fisheries*. San Diego, CA: Academic Press, 177–236.

Bouckaert, E. K., Auer, N. A., Roseman, E. F., and Boase, J. (2014). Verifying success of artificial reefs in the Huron-Erie Corridor for lake sturgeon. *Journal of Applied Ichthyology*, *30*(6): 1393–1401.

Brickhill, M. J., Lee, S. Y., and Connolly, R. M. (2005). Fishes associated with artificial reefs: Attributing changes to attraction or production using novel approaches. *Journal of Fish Biology*, *67*: 53–71.

Carr, M. H., and Hixon, M. A. (1997). Artificial reefs: The importance of comparisons with natural reefs. *Fisheries*, *22*: 28–33.

Claudet, J., and Pelletier, D. (2004). Marine protected areas and artificial reefs: A review of the interactions between management and scientific studies. *Aquatic Living Resources*, *17*: 129–138.

Collier, J. S., and Brown, C. J. (2004). Correlation of sidescan backscatter with grain size distribution of surficial seabed sediments. *Marine Geology*, *214*: 431–449.

Creque, S. M., Raffenberg, M. J., Brofka, W. A., and Dettmers, J. M. (2006). If you build it, will they come? Fish and angler use at a freshwater artificial reef. *North American Journal of Fisheries Management*, *26*: 702–713.

Dumont, P., Amours, J. D., Thibodeau, S., Dubuc, N., Verdon, R., Garceau, S., Bilodeau, P., Mailhot, Y., and Fortin, R. (2011). Effects of the development of a newly created spawning ground in the Des Prairies River (Quebec, Canada) on the reproductive success of lake sturgeon (Acipenser fulvescens). *Journal of Applied Ichthyology*, *27*: 394–404.

Dustin, D. L., and Jacobson, P. C. (2003). Evaluation of walleye spawning habitat improvement projects in streams. *Minnesota Department of Natural Resources, Investigational Report*, *502*: 1–19.

Edsall, T. A., Brown, C. L., Kennedy, G. W., and Poe, T. P. (1992). Lake trout spawning habitat in the Six Fathom Bank-Yankee Reef Trout Sanctuary, Lake Huron. *Journal of Great Lakes Research*, *18*: 70–90.

Edsall, T. A., Poe, T. P., Nester, R. T., and Brown, C. L. (1989). Side-scan sonar mapping of lake trout spawning habitat in northern Lake Michigan. *North American Journal of Fisheries Management*, 9: 269–279.

Environment Illimité Inc. (2009). *Investigations of Lake Sturgeon Spawning Activities at Iroquois Dam on the St. Lawrence River in 2008 – Final Report*. Prepared for the New York Power Authority, White Plains, NY.

Fischer, J. L., Bennion, D., Roseman, E. F., and Manny, B. A. (2015). Validation of a spatial model used to locate fish spawning reef construction sites in the St. Clair-Detroit River System. *Journal of Great Lakes Research*, *41*: 1178–1184.

Fitzsimons, J.D. (1996). The significance of man-made structures for lake trout spawning in the Great Lakes: Are they a viable alternative to natural reefs? *Canadian Journal of Fisheries and Aquatic Sciences*, *53*: 142–151.

Flopp, H., Lowry, M., Gregson, M., and Suthers, I. (2013). Fish assemblages on estuarine artificial reefs: Natural rocky-reef mimics or discrete assemblages? *PLOS ONE*, 8: e63505.

Foster, N. R., and Kennedy, G. W. (1995). Evaluation of Tawas artificial reef as fish spawning habitat. In: Kelso, J. R. M. and Hartig, J. H. (Eds.), *Methods of Modifying Habitat to Benefit the Great Lakes Ecosystem*. Ottawa: CISTI, Occasional Paper No. 1: 267–271.

Gannon, J. (1990). International position statement and evaluation guidelines for artificial reef development in the Great Lakes. *Great Lakes Fishery Commission, Special Publication*: 90–92.

Geiling, W. D., Kelso, J. R. M., and Iwachewski, E. (1996). Benefits from incremental additions to walleye spawning habitat in the Current River, with reference to habitat modification as a walleye management tool in Ontario. *Canadian Journal of Fisheries and Aquatic Sciences*, *53*: 79–87.

Harriott, V. J. and Fisk, D. A. (1988). Coral transplantation as a reef management option. *Proceedings of the 6th International Coral Reef Symposium*, *2*: 375–379.

Hartig, J. H., Zarull, M. A., Ciborowski, J. J. H., Gannon, J. E., Wilke, E., Norwood, G., and Vincent, A. N. (2009). Long-term ecosystem monitoring and assessment of the Detroit River and Western Lake Erie. *Environmental Monitoring and Assessment*, *158*: 87–104.

Hotchkiss, R. H., and Huang, X. (1995). Hydrosuction sediment-removal systems (HSRS): Principles and field test. *Journal of Hydraulic Engineering*, *121*(6): 479–489.

Jelks, H. L., Walsh, S. J., Burkhead, N. M., Contreras-Balderas, S., Diaz-Pardo, E., Hendrickson, D. A., Lyons, J., Mandrak, N. E., McCormick, F., Nelson, J. S., Platania, S. P., Porter, B. A., Renaud, C. B,. Schmitter-Soto, J. J., Taylor, E. B., and Warren. M. L. (2008). Conservation status of imperiled North American freshwater and diadromous fishes. *Fisheries*, *33*: 372–407.

Johnson, J. H., LaPan, S. R., Klindt, R. M., and Schiavone, A. (2006). Lake sturgeon spawning on artificial habitat in the St. Lawrence River. *Journal of Applied Ichthyology*, *22*: 465–470.

Kelch, D. O. (1997). Locating and using Lake Erie's artificial reefs. *Ohio Sea Grant College Program*, FS-072.

Kelch, D. O. (2012). Lake Erie's artificial reef program: Build it and they will come. *Ohio Sea Grant Fact Sheet Series*, FS-072.

Kelch, D. O., Snyder, F. L., and Reutter, J. M. (1999). Artificial reefs in Lake Erie: Biological impacts of habitat alteration. *American Fisheries Society Symposium*, *22*: 335–337.

Kevern, N. R., Biener, W. E., VanDerLaan, S. R., and Cornelius, S. D. (1985). Preliminary evaluation of an artificial reef as a fishery management strategy in Lake Michigan. In: D'Itri, F. M. (Ed.), *Artificial Reefs: Marine and Freshwater Applications*. Chelsea, MI: Lewis Publishers, 443–458.

Kinzel, P. J., Nelson, J. M., Kennedy, G., and Bennion, D. (2016). Use of repeat surveys and flow and sediment transport modeling to support fish spawning reef placement in the Detroit River, MI. In: Constantinescu, G., Garcia, M., and Hanes, D. (Eds.), *River Flow 2016*. London: Taylor and Francis, 2047–2054.

Koeck, B., Tessier, A., Brind'Amout, A., Pastor, J., Bijaoui, B., Dalias, N., Astruch, P., Saragoni, G., and Lenfant, P. (2014). Functional differences between fish communities on artificial and natural reefs: A case study along the French Catalan coast. *Aquatic Biology*, *20*: 219–234.

Kuhl, D. (1992). 14 years of artificial grain feeding in the Rhine downstream of the barrage Iffezheim. In: Larsen, P., and Eisenhauer, N. (Eds.), *Proceedings of the 5th International Symposium on River Sedimentation*. Karlsruhe, Germany: Karlsruhe Institute of Technology, 1121–1129.

Lindahl, U. (2003). Coral reef rehabilitation through transplantation of staghorn corals: Effects of artificial stabilization and mechanical damages. *Coral Reefs*, *22*(3): 217–223.

Lindberg, W. J., and Seaman, W. (ed.) (2011). *Guidelines and Management Practices for Artificial Reef Siting, Use, Construction, and Anchoring in Southeast Florida*. Miami, FL: Florida Department of Environmental Protection.

Manny, B. A. (2006). *Monitoring Element of the Belle Isle/Detroit River Sturgeon Habitat Restoration, Monitoring, and Education Project*. Research Completion Report to Michigan Sea Grant Program, University of Michigan, Ann Arbor, MI.

Manny, B. A., Kennedy, G. W., Boase, J. C., Allen, J. D., and Roseman, E. F. (2010). Spawning by Walleye (Sander vitreus) and white sucker (Catostomus commersoni) in the Detroit River: Implications for spawning habitat enhancement. *Journal of Great Lakes Research*, *36*: 490–496.

Manny, B. A., Roseman, E. F., Kennedy, G., Boase, J. C., Craig, J. M., Bennion, D. H., Read, J., Vaccaro, L., Chiotti, J., Drouin, R., and Ellison, R. (2015). A scientific basis for restoring fish spawning habitat in the St. Clair and Detroit Rivers of the Laurentian Great Lakes. *Restoration Ecology*, *23*(2): 149–156.

Marranca, J. M., Welsh, A., and Roseman, E. F. (2015). Genetic effects of habitat restoration in the Laurentian Great Lakes: An assessment of lake sturgeon origin and genetic diversity. *Restoration Ecology*, *23*(4): 455–464.

Marsden, J. E., Binder, T. R., Johnson, J., He, J., Dingledine, N., Adams, J., Johnson, N. S., Buchinger, T. J., and Krueger, C. C. (2016). Five-year evaluation of habitat remediation in Thunder Bay, Lake Huron: Comparison of constructed reef characteristics that attract spawning lake trout. *Fisheries Research*, *183*: 275–286.

Marsden, J. E., and Chotkowski, M. A. (2001). Lake trout spawning on artificial reefs and the effect of zebra mussels: Fatal attraction? *Journal of Great Lakes Research*, *27*: 33–43.

McLean, M. W., Roseman, E. F., Pritt, J., Manny, B. A., and Kennedy, G. (2014). Overview of artificial reefs in the Laurentian Great Lakes. *Journal of Great Lakes Research*, *41*: 1–8.

Merz, J. E., and Setka, J. D. (2004). Evaluation of a spawning habitat enhancement site for chinook salmon in a regulated California river. *North American Journal of Fisheries Management*, *24*(2): 397–407.

Munday, P. L. (2004). Habitat loss, resource specialization, and extinction on coral reefs. *Global Change Biology*, *10*: 1642–1647.

Murray, J. D. (1994). A policy and management assessment of U.S. artificial reef programs. *Bulletin of Marine Science*, *55*: 960–969.

National Oceanic and Atmospheric Administration (NOAA). (2007). *National Artificial Reef Plan (as amended): Guidelines for Siting, Construction, Development, and Assessment of Artificial Reefs*. U.S. Department of Commerce, Washington, DC.

Nelson, R. W., Dwyer, J. R., and Greenberg, W. E. (1987). Regulated flushing in a gravel-bed river for channel habitat maintenance: A Trinity River fisheries case study. *Environmental Management*, *11*(4): 479–493.

Nichols, S. J., Kennedy, G., Crawford, E., Allen, J., French, J., Black, G., Blouin, M., Hickey, J., Chernyak, S., Hass, R., and Thomas, M. (2003). Assessment of lake sturgeon spawning efforts in the lower St. Clair River, Michigan. *Journal of Great Lakes Research*, *29*: 383–391.

Palmer, M. A., Bernhardt, E. S., Allan, J. D., Lake, P. S., Alexander, G., Brooks, S., Carr, J., Clayton, S., Dahm, C. N., Follstad Shah, J., Galat, D .L., Loss, S. G., Goodwin, P., Hart, D. D., Hassett, B., Jenkinson, R., Kondolf, G. M., Lave, R., Meyer, J. L., O'Donnell, T. K., Pagano, L., and Sudduth, E. (2005). Standards for ecologically successful river restoration. *Journal of Applied Ecology*, *42*: 208–217.

Peck, J. W. (1986). Dynamics of reproduction by hatchery lake trout on a man-made spawning reef. *Journal of Great Lakes Research*, *12*: 293–303.

Perkol-Finkel, S., Shashar, N., and Benayahu, Y. (2006). Can artificial reefs mimic natural reef communities? The roles of structural features and age. *Marine Environmental Research*, *61*: 121–135.

Pickering, H., and Whitmarsh, D. (1997). Artificial reefs and fisheries exploitation: A review of the "attraction versus production" debate, the influence of design and its significance for policy. *Fisheries Research*, *31*: 39–59.

Powels, H., Bradford, M. J., Bradford, R. G., Doubleday, W. G., Innes, S., and Levings, C. D. (2000). Assessing and protecting endangered marine species. *ICES Journal of Marine Science*, *57*: 669–676.

Roseman, E. F., Manny, B., Boase, J., Child, M., Kennedy, G., Craig, J., Soper, K., and Drouin, R. (2011). Lake Sturgeon responses to a spawning reef constructed in the Detroit River. *Journal of Applied Ichthyology*, *27*: 66–67.

Rutecki, T. L., Dorr, J. A., and Jude, D. J. (1985). Preliminary analysis of colonization and succession of selected algae, invertebrates, and fish on two artificial reefs in inshore southeastern Lake Michigan. In: D'Itri, F. M. (Ed.), *Artificial Reefs: Marine and Freshwater Applications*. Chelsea, MI: Lewis Publishers, 459–489.

Rutherford, E. S., Marshall, E., Clapp, D., Horns, W., Gorenflo, T., and Trudeau, T. (2004). *Lake Michigan Environmental Objectives*. Great Lakes Fishery Commission, 2004 Project Completion Report.

Schreiner, D. R., Bronte, C. R., Payne, N. R., Fitzsimons, J. D., and Casselman, J. M. (1995). Use of egg traps to investigate lake trout spawning in the Great Lakes. *Journal of Great Lakes Research*, *21*: 433–444.

Sheng, P. Y. (2000). Physical characteristics and engineering at reef sites. In Seaman, W., and Jensen, A. C. (Eds.), *Artificial Reef Evaluation: With Application to Natural Marine Habitats*. Boca Raton, FL: CRC Press, 51–94.

Smokorowski, K. E., Stoneman, M. G., Cairns, V. W., Minns, C. K., Randall, R. G., and Valere, B. (1998b). Trends in the nearshore fish community of Hamilton Harbour, 1988 to 1997, as measured using an Index of Biotic Integrity. *Canadian Technical Report of Fisheries and Aquatic Sciences*, *2230*: IV–86.

Smokorowski, K. E., Withers, K. J., and Kelso, J. R. M. (1998a). Does habitat creation contribute to management goals? An evaluation of literature documenting freshwater habitat rehabilitation or enhancement projects. *Canadian Technical Report of Fisheries and Aquatic Sciences*, *2249*: v–72.

Spieler, R. E., Gilliam, D. S., and Sherman, R. L. (2001). Artificial substrate and coral reef restoration: What do we need to know to know what we need. *Bulletin of Marine Science*, 69(2): 1013–1030.

Wilding, T. A., and Sayer, M. D. J. (2002). Evaluating artificial reef performance: Approaches to pre- and post-deployment research. *Journal of Marine Science*, *59*: S222–S230.

Part II

Extinction and survival

5 Georgian Bay, Lake Huron

Freshwater turtles and their wetland habitat in a changing landscape

Chantel E. Markle and *Patricia Chow-Fraser*

Introduction

Georgian Bay, the eastern arm of Lake Huron, is home of the world's largest freshwater archipelago, a World Biosphere Reserve (UNESCO, 2014) that contains more than 30,000 islands (Figure 5.1). Considered the sixth Great Lake by some (Barry, 1978), Georgian Bay spans 15,000 square kilometers (5,800 square miles) and is almost as large as Lake Ontario (18,960 square kilometers [7,320 square

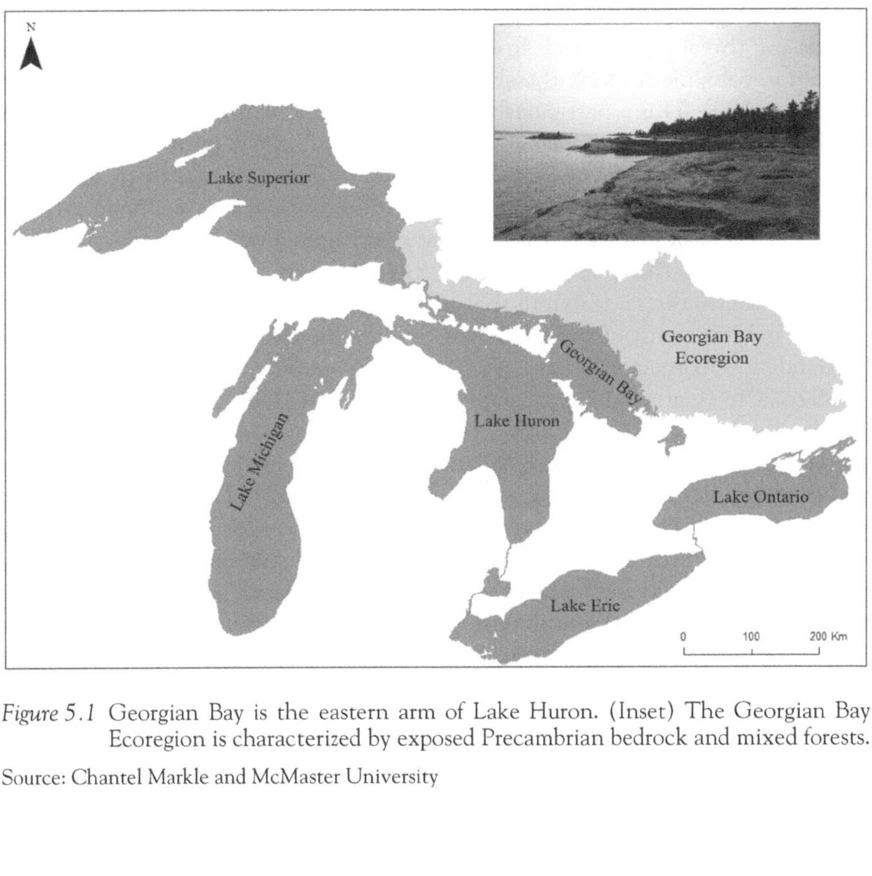

Figure 5.1 Georgian Bay is the eastern arm of Lake Huron. (Inset) The Georgian Bay Ecoregion is characterized by exposed Precambrian bedrock and mixed forests.

Source: Chantel Markle and McMaster University

miles]). Unlike Lake Ontario, Lake Erie, and Lake Michigan, which are underlain by sedimentary rock (sandstone and limestone), bedrock in Georgian Bay is Canadian Shield, which is granitic with only a very thin layer of soil (Georgian Bay Ecoregion; Figure 5.1). This bedrock geology, together with exposure to prevailing winds, has inspired the iconic image of windswept pines on top of rocky shorelines popularized by Canada's Group of Seven artists. Georgian Bay is also unique because its convoluted shoreline has led to the development of many small coastal wetlands. In fact, more than 12,600 distinct wetlands have been identified within the bay's coastal zone (Midwood et al., 2012). This is more than 24 times as many coastal wetlands as Lake Michigan (524), Lake Huron (493), Lake Ontario (488), Lake Superior (307), and Lake Erie (149) (Ingram et al., 2004). Although most of the wetlands in Georgian Bay are small (< 2 hectares [0.008 square miles]), they tend to form complexes that function like larger wetlands, providing reproductive and foraging habitat for many fish species and aquatic wildlife (Midwood and Chow-Fraser, 2014). In addition to having the highest density of coastal wetlands in the Great Lakes Basin, it has a disproportionate number of contiguous pristine wetlands (Cvetkovic and Chow-Fraser, 2011).

In a previous study, only 2 of the 175 coastal wetlands surveyed in eastern Georgian Bay showed signs of degradation attributable to human activities per the Water Quality Index (Chow-Fraser, 2006). Although the remaining 99 percent of surveyed coastal wetlands are in good to excellent condition, there are increasing threats to wetlands from road expansion, as well as cottage and residential development along the shoreline (Niemi et al., 2007; Walton and Villeneuve, 1999). This is particularly true in southern Georgian Bay, where coastal wetlands have been developed at the expense of wildlife habitat. Since southern Georgian Bay is only a two-hour drive north of Toronto, it is easily accessible to weekend users, making it a popular vacation destination. Georgian Bay also contains the busiest recreational waterway in Canada (Walton and Villeneuve, 1999). These factors put species and their habitats at risk (Bywater, 2013), although not to the same extent that urbanization and agricultural development impact habitat south of the Canadian Shield. Another factor affecting coastal wetlands in Georgian Bay has been 15 years of sustained low water levels, from 1999 to 2014 (Canadian Hydrographic Service, 2012); only recently have water levels rebounded above the long-term average. The loss of interannual variation of water levels significantly affects the structure of vegetation communities in coastal wetlands (Midwood and Chow-Fraser, 2012). The loss of vegetation structure and diversity can have cascading effects on species reliant on coastal wetlands. Given the unprecedented nature of this disturbance, we do not know how long it will take before coastal wetlands will recover, or indeed, if they will transition to an altogether different state, especially with the changing climate. All these threats could fundamentally change ecosystem processes in coastal marshes of Georgian Bay, and we have the rare opportunity to study them as they adapt to new conditions.

These coastal and upland wetlands provide critical breeding, foraging, and overwintering habitat for many rare avian, fish, and reptile species, and it is well documented that degradation in wetland quality affects their abundance (DeCatanzaro and Chow-Fraser, 2010; Seilheimer et al., 2007; Smith and Chow-Fraser, 2010). A good example is the 33 species of reptiles and amphibians that live in the Georgian Bay archipelago, a higher herpetofauna diversity than anywhere else in Canada (Parks Canada, 2013). Of these species, the coastal zone of Georgian Bay supports six species of freshwater turtles: the spotted turtle (*Clemmys guttata*), snapping turtle (*Chelydra serpentina*), midland painted turtle (*Chrysemys picta marginata*), northern map turtle (*Graptemys geographica*), eastern musk turtle (*Sternotherus odoratus*), and Blanding's turtle (*Emydoidea blandingii*). All except the painted turtle are listed as species at risk in this region (Box 5.1).

Box 5.1 What is a species at risk?

A species at risk is a species facing population declines or whose habitat faces threats. Typically, this is due to human-based impacts. The species may become extinct if nothing is done to reverse these threats.

Although turtles are one of the oldest species in the world and have survived ice ages and mass extinctions, habitat loss and degradation following European settlement over the past 200 years have led to their current decline (Ernst and Lovich, 2009). In particular, the Blanding's turtle is listed as a threatened species provincially and endangered federally in its Ontario Great Lakes range due to habitat alteration and destruction (Government of Canada, 2017; Marchand and Litvaitis, 2004; Ontario Government, 2007 Steen and Gibbs, 2004). The Blanding's turtle is an important focal species because it is a semi-aquatic species known to make extensive use of both aquatic and terrestrial habitats. Across the species' range, Blanding's turtles are known to use aquatic habitats such as vernal pools, bogs, marshes, and fens (Edge et al., 2010; Hartwig and Kiviat, 2007; Markle and Chow-Fraser, 2014) and terrestrial habitats such as forests throughout the active season (Ernst and Lovich, 2009). Their use of habitat mosaics makes them the perfect candidate for landscape-level conservation approaches (Box 5.2). Furthermore, they can be thought of as an umbrella species, which means that if we can identify and protect habitats required for the Blanding's turtle, we will likely protect many other species in the process. This is especially beneficial in Georgian Bay because it is difficult to carry out extensive field sampling for many species in remote wetlands. By designing a detailed landscape conservation plan for an umbrella species, we can better direct resources while having the greatest positive impact.

Box 5.2 Landscape ecology and conservation

Landscape ecology is the study of spatial patterns and their interactions at a variety of spatial scales. Research may involve mapping or analysis of patterns or habitat structure in different landscapes, their change through time, and the impact of human disturbance. Studies may also examine relationships and interactions among landscape features and habitat composition and its species. Conservation strategies at the landscape scale are designed to consider a more holistic approach to managing a particular area and are generally applied at a large scale.

Statement of the problem

In response to increased pressure from human development, which threatens freshwater turtles and their habitats in this region, we conducted research to enhance conservation strategies for Blanding's turtles in the Georgian Bay archipelago. The geologically distinct landscape that dominates the coastal islands here make it difficult for us to directly apply information to our study sites from populations living elsewhere in Canada and the United States. (e.g., Beaudry et al., 2009; Edge et al., 2009, 2010; Hartwig and Kiviat, 2007; Millar and Blouin-Demers, 2011). When designing landscape approaches, it's important to apply them in the appropriate setting. Therefore, we collected data to develop a habitat suitability model specific to the Georgian Bay landscape. We developed the model using information collected from a population of Blanding's turtles on a protected island and then applied our results to the entire archipelago to produce regional maps for conservation and management purposes. To track the types of habitat used by turtles during various activities (e.g., mating, feeding, nesting, and overwintering, etc.), we used radio telemetry techniques to follow turtles for more than two years. This interdisciplinary approach combined field data, remote sensing, and statistical modeling to produce spatially-explicit statistical models that should advance efforts to develop effective management plans for Blanding's turtles throughout the biosphere reserve. The overall goal is to identify all suitable islands in the archipelago to mark them for protection in conservation plans before they become degraded or developed.

Methods

Radio tracking

Radio tracking allows us to locate a tagged turtle throughout the season to study movement patterns and habitat use. At the beginning of the study, we attached a radio transmitter to the back of each turtle's shell with a quick dry epoxy (Figure 5.2). We ensured the transmitter had a streamlined finish to allow the turtle to move easily through water and vegetation. We then used a receiver

Figure 5.2 A radio transmitter is attached to a turtle's shell with epoxy.

Source: Chantel Markle

and antenna to scan for the unique radio frequency emitted by each transmitter (Figure 5.3). We radio tracked turtles at least once per week during the active season and approximately once a month during the winter. Whenever a turtle was located, we recorded the date, time of day, GPS location, turtle's activity, weather

Figure 5.3 A researcher uses a receiver and antenna to detect the signal from a radio transmitter and locate a tagged turtle.

Source: Chantel Markle

conditions, and details about the habitat. We used the series of radio locations for each turtle to determine where the turtle moved throughout the year and the key habitats it used.

GPS logger

In some cases, we also attached a GPS logger to a turtle to collect additional locational data on a predetermined schedule (Figure 5.4). By changing the timing of when the GPS attempted to record a location, we could increase (e.g., every hour to capture small-scale movement) or decrease the number of fix attempts (e.g., every few days to capture large-scale movement). This is useful for collecting data on female nesting movements that tend to occur at night when researchers are generally not in the field to radio track a turtle. GPS loggers are an important tool to enable researchers to accurately map the extent of turtle movements (Christensen and Chow-Fraser, 2014).

Geographic information systems

We used a geographic information system (GIS) to display and analyze our spatial data. In a GIS, we created maps of our study area, displayed turtle locations and

Figure 5.4 A GPS logger is attached with epoxy. Pictured is a GPS logger without a radio transmitter built in, so the turtle is also outfitted with a transmitter.

Source: Chantel Markle

movements, and mapped suitable habitats. In addition, we used aerial imagery of our study site as a wayto see the entire study area and map out the location of important habitats, such as wetlands and forests. Depending on the time of year the aerial image was acquired, we could also determine the amount of vegetation that has grown in an area of interest. By using unmanned aerial vehicles (i.e., drones) we can capture high-resolution aerial images at the most desired time, even multiple times a season, or annually for several years to quantify habitat change. Unmanned aerial vehicles are especially useful in dynamic systems such as coastal wetlands where the percent of floating and emergent vegetation fluctuates throughout the summer and requires current aerial images to produce accurate results (Marcaccio et al., 2016).

Findings

Habitat use

Blanding's turtles use a mosaic of aquatic and terrestrial habitats, exhibiting fidelity to their residence wetlands, and returning year after year to the same wetland (Congdon et al., 2011). Generally, they emerge in April or May from overwintering habitats, such as permanent pools (Joyal et al., 2001; Ross and Anderson, 1990), streams (Newton and Herman, 2009), and wetlands (Edge et al., 2009; Kofron and Schreiber, 1985). During the spring, turtles often bask on logs or mats of vegetation to increase their body temperature. Turtles may also move among wetlands in search of food, while males additionally search for mates. In early summer, females begin their journey to nesting sites, and this can include sojourns across multiple wetland types, such as vernal pools, bogs, and marshes (e.g., Beaudry et al., 2009; Markle and Chow-Fraser, 2014; Standing et al., 1999). In addition to diverse habitat use, males and females may move long distances to mate with adjacent subpopulations to increase gene flow (McGuire et al., 2013). Females have been known to move more than 6 kilometers (3.73 miles) in search of a nesting site (Edge et al., 2010), and we found males moving 900 meters (0.56 miles) in early summer in Georgian Bay (Markle and Chow-Fraser, 2014).

Our research highlights the extent and variety of habitat types used by Blanding's turtles in the Georgian Bay archipelago to carry out their life processes (Markle and Chow-Fraser, 2014; Figure 5.5). Initially, females used inland wetlands for feeding and basking. Two bogs were resident wetlands for all tagged turtles (i.e., bearing transmitters; n = 15) and were dominated by 30 to 38 percent cover of *Sphagnum* moss mats. In late May, females used patches of swamp and isolated vernal pools to move to coastal wetlands – considered staging habitat – where they spent one to two weeks feeding before migrating to nesting grounds. Females nested in soil-filled crevices in open rocky outcrops, a nesting site type unique to the Canadian Shield. In southwestern Ontario, majority of nesting occurs along sandy beaches or roadsides in urbanized areas. Upon completion of nesting, females traveled about 400 meters (0.25 miles) to return to their resident wetlands where all tagged turtles successfully overwintered. The males' habitat use and movements were less extensive than those of females. Over the course of the spring and summer, males spent most of their time in two bogs, using a chain of small vernal pools to move among them. Our study identified

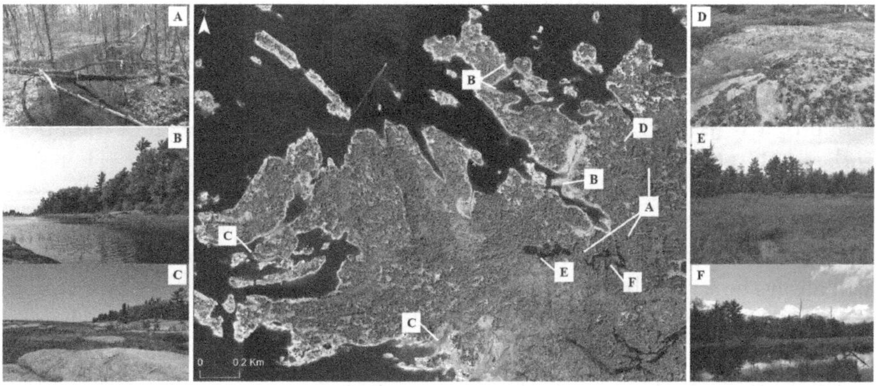

Figure 5.5 A population of Blanding's turtles inhabited approximately 293 hectares (1.13 square miles) throughout the year. Turtles used vernal pools (A) to move among key habitats, including coastal wetlands (B and C), nesting habitat (D), and residence wetlands (E and F), where all turtles overwintered.

Source: Chantel Markle and McMaster University

vernal pools – seasonally inundated wetlands – as a critical habitat type for Blanding's turtles in Georgian Bay. Both males and females used them as corridors to access permanent wetlands and nesting sites. Since vernal pools are sourced by snowmelt and spring rainwater, they begin to dry up through the summer before refilling in fall. However, this cycle of filling and draining depends on weather patterns. In years with reduced snowmelt and warmer winter temperatures, there was reduced distribution of vernal pools throughout the landscape, and this apparently resulted in turtles using other connecting corridors. Climate change is expected to influence the abundance and accessibility of vernal pools, but the extent to which this may alter habitat connectivity and movements of Blanding's turtles is unknown and should be investigated.

Distribution of suitable habitat

By developing a habitat suitability model, we identified the primary habitat requirements for the Blanding's turtle in the Georgian Bay islands. We also created a map of all the islands with suitable habitats for the turtles, and by extension, additional wetland-dependent species. Our resulting maps scored the suitability of an island for Blanding's turtles on a scale of zero to one, and it was statistically derived to reflect the similarity of an island to the habitat composition required by our previously studied subpopulation of turtles. We interpreted an island with a score of zero as having a low probability of containing suitable habitat for Blanding's turtles. On the other end of the scale, a score of one would indicate that the island has a high probability of providing suitable habitat.

Our habitat suitability map indicates that approximately 64 percent of the evaluated islands, or 89 percent (14,027 hectares [54 square miles]) of the total mapped area, in the archipelago is suitable for Blanding's turtles. Our results confirm the assumption that this coastal zone provides high-quality habitats for species at risk. In comparison to the remaining habitat fragments in southwestern Ontario, it is likely that habitats in the archipelago contain much of the undisturbed habitat within the provincial range of the Blanding's turtle. That said, the presence of suitable habitat does not equate to turtle occupancy because of competition for food and other resources, predator abundance, and quality of nesting and overwintering sites. Therefore, the distribution of suitable habitat will always be larger than the actual distribution of the species. For conservation purposes, however, it is better to overestimate the distribution of suitable habitat and include areas with suitable habitat but no turtle population than to risk overlooking a single subpopulation of this at-risk species.

Discussion and conclusions

Within Georgian Bay, freshwater turtles face many threats, including habitat loss and degradation through shoreline modification and cottage development, road mortality, and climate change. Although these same threats exist in other regions, the geological landscape in Georgian Bay is unique compared to other areas within the turtle's range, making it difficult to automatically apply habitat data collected from other populations to the area. Therefore, we collected data to use in developing regional management strategies for Blanding's turtles in the coastal areas of Georgian Bay. One of the first, and perhaps most important, steps when developing a regional strategy is to map suitable habitat for the at-risk species. Next, identifying threats to the population and habitat allows for targeted management strategies. Roads are a concern for many turtle populations because they fragment habitats, and such fragmentation can have a negative impact on species that migrate long distances, often resulting in individuals crossing roads and leaving protected areas. By knowing where suitable habitats exist in relation to roads, land managers and conservation authorities can mitigate road mortality by constructing fences and ecopassages to reconnect fragmented habitat (e.g., Aresco, 2005; Beier and Noss, 1998; Lesbarreres and Fahrig, 2012). Reduction of adult mortality can play a major role in the recovery of declining populations because freshwater turtles have long life spans, delayed sexual maturity, and low rates of juvenile recruitment (Congdon et al., 1993; Marchand and Litvaitis, 2004; Steen and Gibbs, 2004). Therefore, protection of pristine habitat should be the first priority. Subsequently, populations experiencing high adult mortality rates should be identified and threats should be mitigated. Once a source of increased adult mortality is addressed, other conservation measures such as nest protection and head-starting (incubating eggs, raising turtles, and releasing them into the wild when they are larger, giving them a "head-start" at life) can be implemented to improve juvenile recruitment.

As a starting point, Georgian Bay conservation plans should emphasize protection of unique open-rocky nesting habitats; upland and coastal wetlands used for critical activities, such as feeding, mating, and overwintering; and the connecting habitats used to travel among these key areas. Many studies have confirmed the frequent use of wetlands and vernal pools by the Blanding's turtles throughout the active season (Joyal et al., 2001; Millar and Blouin-Demers, 2011) and the importance of wetland areas in relation to increasing turtle occupancy (Fortin et al., 2012). Other modeling work indicated that the risk of turtle extinction increased as vernal pools were lost from the landscape (Gibbs, 1993). Similarly, we found that vernal pools are important features in the Georgian Bay landscape with high ecological value within this ecosystem. We also found that use of vernal pools as a connecting corridor varied among years, and we hypothesize that variation in annual precipitation may affect the timing of turtle movements and use of temporary habitats. In the face of climate change, the spatial distribution of vernal pools is likely to change and is expected to impact the timing of turtle movement patterns throughout the landscape. There is an urgent need for natural resource managers to be aware of this because vernal pools receive no formal protection under the law as an independent category by provincial agencies. Currently, the only way to protect vernal pools is to classify each one, on a case-by-case basis, as part of a wetland complex through the Ontario Wetland Evaluation System (Ontario Ministry of Natural Resources and Forestry, 2013b) or as habitat under the Blanding's turtle habitat regulation (Ontario Ministry of Natural Resources and Forestry, 2013a). Future research should aim to understand how a changing climate will affect the distribution of seasonal wetlands across the landscape.

Although most of coastal Georgian Bay and the archipelago has limited human disturbance, some of islands are being threatened by increasing development pressures. Therefore, ensuring valuable habitats are protected is essential for long-term conservation efforts. For areas experiencing higher levels of development, such as in Severn Sound and Honey Harbour, creating local habitat suitability maps can help guide more detailed management plans. Now that we have begun identifying suitable habitats, it is critical to begin a strategic monitoring program to survey the actual distribution of Blanding's turtles in this area. A lot of work remains if we intend to protect the turtles and high-quality habitat from certain degradation and fragmentation. Ensuring that the landscape provides the habitats required to sustain a population of any species at risk should be the most important aspect of any conservation or mitigation strategy. Without the habitat that species require, no amount of population augmentation, nest protection, and road fencing will ensure long-term population persistence.

Acknowledgements

We would like to thank the students from the Chow-Fraser lab who participated in the field work. We are grateful to those who provided logistical support throughout the projects. Funding was provided in part by the Sierra Club Canada

Foundation, an Ontario Graduate Scholarship, and a CGS-D NSERC. All studies were carried out under the approved McMaster University Animal Use Protocols, government permits, and local site permits.

References

Aresco, M. J. (2005). The effect of sex-specific terrestrial movements and roads on the sex ratio of freshwater turtles. *Biological Conservation*, *123*: 37–44.

Barry, J. P. (1978). *Georgian Bay: The Sixth Great Lake*. Toronto: Clarke, Irwin, p. 190.

Beaudry, F., deMaynadier, P. G., and Hunter, M. L. (2009). Seasonally dynamic habitat use by spotted (*Clemmys guttata*) and Blanding's turtles (*Emydoidea blandingii*) in Maine. *Journal of Herpetology*, 43(4): 636–645.

Beier, P., and Noss, R. F. (1998). Do habitat corridors provide connectivity? *Conservation Biology*, *12*(6): 1241–1252.

Bywater, D. (2013). State of the Bay background report. *Georgian Bay Biosphere Reserve*: 190.

Canadian Hydrographic Service. (2012). *Historical Water Level Data*. Available at www.waterlevels.gc.ca/C&A/tidal_e.html

Chow-Fraser, P. (2006). Development of the wetland water quality index for assessing the quality of Great Lakes coastal wetlands. *Coastal Wetlands of the Laurentian Great Lakes: Health, Habitat and Indicators*, 5: 137–166.

Christensen, R. J., and Chow-Fraser, P. (2014). Use of GPS loggers to enhance radio-tracking studies of freshwater turtles. *Herpetological Conservation and Biology*, 9(1): 18–28.

Congdon, J. D., Dunham, A. E., and van Loben Sels, R. C. (1993). Delayed sexual maturity and demographics of Blanding's turtles (*Emydoidea blandingii*): Implications for conservation and management of long-lived organisms. *Journal of Conservation Biology*, *7*: 826–833. doi: 10.1046/j.1523–1739.1993.740826.x

Congdon, J. D., Kinney, O. M., and Nagle, R. D. (2011). Spatial ecology and core-area protection of Blanding's Turtle (*Emydoidea blandingii*). *Canadian Journal of Zoology*, 89(11): 1098–1106.

COSEWIC. (2005). *COSEWIC Assessment and Update Status Report on the Blanding's Turtle (Emydoidea blandingii) in Canada*. Ottawa: Committee on the Status of Endangered Wildlife in Canada.

Cvetkovic, M., and Chow-Fraser, P. (2011). Use of ecological indicators to assess the quality of Great Lakes coastal wetlands. *Ecological Indicators*, *11*: 1609–1622.

DeCatanzaro, R., and Chow-Fraser, P. (2010). Relationship of road density and marsh condition to turtle assemblage characteristics in the Laurentian Great Lakes. *Journal of Great Lakes Research*, 36: 357–365.

Edge, C. B., Steinberg, B. D., Brooks, R. J., and Litzgus, J. D. (2009). Temperature and site selection by Blanding's turtles (*Emydoidea blandingii*) during hibernation near the species' northern range limit. *Canadian Journal of Zoology*, 87(9): 825–834. doi: 10.1139/z09–073

Edge, C. B., Steinberg, B. D., Brooks, R. J., and Litzgus, J. D. (2010). Habitat selection by Blanding's turtles (*Emydoidea blandingii*) in a relatively pristine landscape. *Ecoscience*, *17*(1): 90–99. doi: 10.2980/17–1–3317

Ernst, C. H., and Lovich, J. E. (2009). *Turtles of the United States and Canada* (2nd ed.). Baltimore, MD: Johns Hopkins University Press.

Fortin, G., Blouin-Demers, G., and Dubois, Y. (2012). Landscape composition weakly affects home range size in Blanding's turtles (*Emydoidea blandingii*). *Ecoscience*, 19(3): 191–197.

Gibbs, J. P. (1993). Importance of small wetlands for the persistence of local populations of wetland-associated animals. *Wetlands*, *13*(1): 25–31.

Government of Canada. (2017). *Blanding's Turtle Great Lakes/St. Lawrence population.* Species at Risk Public Registry. Available at: www.registrelep-sararegistry.gc.ca.

Hartwig, T., and Kiviat, E. (2007). Microhabitat association of Blanding's turtles in natural and constructed wetlands in southeastern New York. *The Journal of Wildlife Management*, *71*(2): 576–582.

Ingram, J., Holmes, K., Grabas, G., Watton, P., Potter, B., Gomer, T., and Stow, N. (2004). Development of a coastal wetlands database for the Great Lakes Canadian shoreline. *Final Report to the Great Lakes Commission (Wetlands2-EPA-03)*, *51*: 18.

Joyal, L. A., McCollough, M., and Hunter, M. L. (2001). Landscape ecology approaches to wetland species conservation: A case study of two turtle species in southern Maine. *Conservation Biology*, *15*(6): 1755–1762. doi: 10.1046/j.1523–1739.2001.98574.x

Kofron, C. P., and Schreiber, A. A. (1985). Ecology of two endangered aquatic turtles in Missouri: *Kinosternon flavescens* and *Emydoidea blandingii*. *Journal of Herpetology*, *19*(1): 27–40.

Lesbarreres, D., and Fahrig, L. (2012). Measures to reduce population fragmentation by roads: What has worked and how do we know? *Trends in Ecology & Evolution*, *27*(7): 374–380.

Marcaccio, J. V., Markle, C. E., and Chow-Fraser, P. (2016). Use of fixed-wing and multi-rotor unmanned aerial vehicles to map dynamic changes in a freshwater marsh. *Journal of Unmanned Vehicle Systems*, *4*(3): 193–202.

Marchand, M. N., and Litvaitis, J. A. (2004). Effects of habitat features and landscape composition on the population structure of a common aquatic turtle in a region undergoing rapid development. *Conservation Biology*, *18*(3): 758–767. doi: 10.1111/j.1523- 1739.2004.00019.x

Markle, C. E., and Chow-Fraser, P. (2014). Habitat selection by the Blanding's turtle (*Emydoidea blandingii*) on a protected island in Georgian Bay, Lake Huron. *Chelonian Conservation and Biology*, *13*(2): 216–226.

McGuire, J. M., Scribner, K. T., and Congdon, J. D. (2013). Spatial aspects of movements, mating patterns, and nest distributions influence gene flow among population subunits of Blanding's turtles (*Emydoidea blandingii*). *Conservation Genetics*, *14*(5): 1029–1042.

Midwood, J. D., and Chow-Fraser, P. (2012). Changes in aquatic vegetation and fish communities following 5 years of sustained low water levels in coastal marshes of eastern Georgian Bay, Lake Huron. *Global Change Biology*, *18*(1): 93–105.

Midwood, J. D., and Chow-Fraser, P. (2014). Connecting coastal marshes using movements of resident and migratory fishes. *Wetlands*, *35*(1): 69–79.

Midwood, J. D., Rokitnicki-Wojcik, D., and Chow-Fraser, P. (2012). Development of an inventory of coastal wetlands for eastern Georgian Bay, Lake Huron. *ISRN Ecology*, *2012*, 1–12. doi: 10.5402/2012/950173

Millar, C. S., and Blouin-Demers, G. (2011). Spatial ecology and seasonal activity of Blanding's turtles (*Emydoidea blandingii*) in Ontario, Canada. *Journal of Herpetology*, *45*(3): 370–378. doi: 10.1670/10–172.1

Newton, E. J., and Herman, T. B. (2009). Habitat, movements, and behaviour of overwintering Blanding's turtles (*Emydoidea blandingii*) in Nova Scotia. *Canadian Journal of Zoology*, *87*(4): 299–309. doi: 10.1139/Z09–014

Niemi, G., Kelly, J., and Danz, N. (2007). Environmental indicators for the coastal region of the North American Great Lakes: Introduction and prospectus. *Journal of Great Lakes Research*, *33*(Special Issue 3): 1–12.

Ontario Government. (2007). *Endangered Species Act*. Available from https://www.ontario.ca/laws/statute/07e06 [Accessed 27 May 2016].

Ontario Ministry of Natural Resources and Forestry. (2013a). *General Habitat Description for the Blanding's Turtle*. Canada: Ontario Ministry of Natural Resources and Forestry. Availble at: https://www.ontario.ca/page/blandings-turtle-general-habitat-description.

Ontario Ministry of Natural Resources and Forestry. (2013b). *Ontario Wetland Evaluation System, Northern Manual*. Ontario Ministry of Natural Resources and Forestry, Peterborough, Ontario. 1st Edition, Version 1.2. p. 288.

Parks Canada. (2013). *Georgian Bay Islands National Park of Canada: Natural Heritage*. Available at www.pc.gc.ca/eng/pn-np/on/georg/natcul/natcul1.aspx

Ross, D. A., and Anderson, R. K. (1990). Habitat use, movements, and nesting of *Emydoidea blandingii* in central Wisconsin. *Journal of Herpetology*, *24*(1): 6–12.

Seilheimer, T., Wei, A., Chow-Fraser, P., and Eyles, N. (2007). Impact of urbanization on the water quality, fish habitat, and fish community of a Lake Ontario marsh, Frenchman's Bay. *Urban Ecosystems*, *10*: 299–319. doi: 10.1007/s11252-007-0028-5

Smith, L., and Chow-Fraser, P. (2010). Impact of adjacent land use and isolation on marsh bird communities. *Environmental Management*, 45: 1040–1051. doi: 10.1007/s00267-010-9475-5

Standing, L., Herman, T., and Morrison, I. (1999). Nesting ecology of Blanding's turtle (*Emydoidea blandingii*) in Nova Scotia, the northeastern limit of the species' range. *Canadian Journal of Zoology*, *77*(10): 1609–1614. doi: 10.1139/z99–122

Steen, D. A., and Gibbs, J. P. (2004). Effects of roads on the structure of freshwater turtle populations. *Conservation Biology*, *18*(4): 1143–1148.

UNESCO. (2014). *Directory of the World Network of Biosphere Reserves*. Available at www.unesco.org/new/en/natural-sciences/environment/ecological- sciences/biosphere-reserves/world-network-wnbr/wnbr/

Walton, M., and Villeneuve, M. (1999). *Ecosystem Planning in Georgian Bay Islands National Park: A Multi-Jurisdictional Approach*. Second international symposium and workshop on the conservation of the eastern massasauga rattlesnake, *Sistrurus catenatus*: population and habitat management issues in urban, bog, prairie and forested ecosystems, Toronto Zoo, Toronto, Ontario.

6 Framing extinction

Societal attitudes toward the passenger pigeon in editorials and opinion pieces

Bruno Takahashi, Ran Duan, Apoorva Joshi, Anthony Van Witsen, and *Eric Freedman*

Introduction

In October 2016, a World Wildlife Fund report warned that about two-thirds of the planet's wildlife could disappear by 2020. The report received considerable media coverage, and although it has been challenged as overstatement, its findings mostly align with much of the scientific research confirming the dramatic impact of human activities on biodiversity loss (Ceballos et al., 2015). Species extinction has always been of interest to various publics, but in the twenty-first century, the impacts of people on extinctions are unprecedented. Some scientists refer to this sixth extinction as an event "as extensive and rapid as five previous mass extinction events in the last 600 million years when 65 percent to 95 percent of marine animals previously known from fossil records disappeared" (Stork, 2009: 58). The current era, referred to as the Anthropocene (Young et al., 2016), raises questions about how modern society approaches the problem. One way to understand societal attitudes about extinction is to examine how we use language to refer to species and their conservation.

The passenger pigeon (*Ectopistes migratorius*), an iconic species, went extinct more than 100 years ago, when the last pigeon, Martha, died in the Cincinnati Zoo in Ohio. Her body was encased in ice and shipped to the Smithsonian Institution in Washington, D.C., where she was mounted and is now kept in a nondescript steel storage locker (Freedman, 2011).

Many news media used the 100th anniversary of Martha's death in 2014 to reflect on ways people impact and conserve nonhuman species. This study presents a qualitative exploration of the frames (Entman, 1993) used by writers of opinion pieces and editorials about this anniversary in print and online publications in the United States and Canada.

Statement of the problem and research question

Attitudes toward species, extinction, and conservation

Views about animals, environmental attitudes, and perceptions about nature's value can explain how we treat nonhuman species and how we approach conservation efforts (de Pinho et al., 2014; George et al., 2016). Attitudes, in and of themselves, are an evaluation and a "tendency to respond" positively or negatively to the object in question (Fishbein, 1975). In this case, the attitude is toward animals broadly, and more specifically toward endangered species and extinction. Although extensive research on attitudes toward animals and conservation has examined anthropomorphic characteristics (Batt, 2009), children's attitudes (Kellert and Westervelt, 1984), animals' utilitarian value such as for food (Knight et al., 2004), and flagship species (Schlegel and Rupf, 2010), attitudes toward animal use are multidimensional and depend on context (Knight et al., 2010). Scientists, for instance, might believe using animals for medical science is a necessary evil benefitting people, while animal rights advocates may focus on morality and the cost to the animals themselves. Justification for protecting a species could also be based on belief in its potential "recreational" value to those who participate in wildlife-watching (Kerlinger et al., 2013) or on the belief that some species symbolize important ideologies, like the association between the bald eagle and American democracy (Petersen, 1999; Raymond and Schneider, 2014; Beans, 1997). Overall, the belief that a species is ecologically valuable because it helps keep the food chain intact, and thus helps maintain the balance of all life, is a strong basis for wanting to protect endangered species (Sax, 2001; Stein et al., 2000).

Media and the coverage of animals and extinction

Attitudes toward animals are manifested through a variety of mediated forms, such as news media (Anderson, 1997). For example, in an analysis of how hunting magazines portrayed the display of trophy animals, Kalof and Fitzgerald (2003) found extreme objectification and marginalization of animal bodies rather than love and respect for nature. Their study also linked collecting and exhibiting wild animals to domination, patriarchy, and colonialism.

Media representations of animals, extinction, and conservation in opinion pieces are important because that form is less constrained by journalistic norms and routines than news articles and has a stronger potential impact on public policy (Sommer and Maycroft, 2008). Op-eds, in particular, give voice to non-staff opinions that are separate – if not necessarily different – from a newspaper editorial board's (Ciofalo and Traverso, 1994; Day and Golan, 2005; Sommer and Maycroft, 2008). Research has focused on characteristics of opinion piece authors (Golan, 2013; Mitman et al., 2012) and their communication strategies, such as rhetorical devices (Parks and Takahashi, 2016). Although such pieces are venues for public deliberation, most research shows only a limited diversity of voices. Instead, editorials, op-eds, and letters were used strategically for the authors' or the newspapers' specific purposes.

Little research has examined frames in opinion pieces on conservation issues. This study builds on previous non-framing research on opinion pieces to examine the moral framing – related to moral concerns of harm and care – of environmental issues (Feinberg and Willer, 2013).

The extinction of the passenger pigeon

Endangered species are often iconic in appeal, prompting some to be labeled flagship species: popular and charismatic species that stand as symbols stimulating conservation awareness and action (Douglas and Verissimo, 2013). This strategy has great potential to change people's behavior, including participating in conservation-related activities and fundraising (Leader-Williams & Dublin, 2000). Habitat demands of flagship species often correspond to needs of other species inhabiting the same regions, thus turning some flagship species into umbrella species (Schlegel and Rupf, 2010). However, not all flagship and iconic species are saved.

To understand social attitudes toward extinction, we focused on an iconic species: the passenger pigeon, which went extinct in 1914. Passenger pigeons once covered the skies of the Great Lakes Basin and elsewhere in North America, with historical accounts suggesting their population was in the billions. As a Smithsonian scientist explained, "One of their downfalls was their social structure. They had to be in huge colonies" (Freedman, 2011).

There is clear evidence that this species was "bewilderingly vast" in 1860 but virtually gone by 1900 (Greenberg, 2014: 192), and that it was forced westward and away from its eastern habitat (Fuller, 2014). Martha's death as an endling – the last known survivor of a species – was a milestone and a reminder of the fragility and vulnerability of seemingly strong and resilient species. Four years later, the last known Carolina parakeet (*Conuropsis carolinensis*) died in the same cage in the same Ohio zoo (Audubon, 2017). The 100th anniversary of Martha's death provided an opportunity to remind us about the importance of conservation (Schulz et al., 2014).

The passenger pigeon was nomadic, traveling in vast numbers, driven more by a need for food than seasonal changes (Greenberg, 2014). It was unclear whether the drop in its population was sudden, whether the population consistently replenished itself for decades in the face of extensive hunting – 50,000 a day were reportedly shot near Petoskey, Michigan, along the Lake Michigan shore in 1878, for example – or whether the numbers dropped dramatically while people remained unaware of the collapse. Fuller (2014) writes that as far as the decline goes, "the logical conclusion must surely be" that numbers had been dropping for years but because there were still millions of birds left, people largely remained indifferent to the plummeting numbers. When the "tipping point" came though, things quickly went south.

Although hunting was deemed primarily responsible for its extinction, that was not the only pressure the pigeons faced and might not have been the most critical. Greenberg (2014: 21) writes, "A number of known factors took their toll on birds, although the principal limitation was the forage base itself. As many birds as there were, the population could not exceed the capacity of their environment to sustain them."

Massive deforestation undermined a species that needed forests for food and shelter, especially because of their vast colonies. Cutting forests made land available for agricultural or residential purposes; elsewhere, habitat was destroyed because trees were needed for firewood and building material. Adverse weather has always affected migratory birds (Greenberg, 2014), while predators, including hawks, owls, martens, weasels, raccoons, black bears, bobcats, and mountain lions, might have preyed on wounded birds. "It is impossible that billions of passenger pigeons did not affect the ecosystems that provided them with sustenance and quarters. Nor is it possible that their elimination in such a short time did not also impact those forests" (Greenberg, 2014: 25).

Thus our research question:

> How do editorial and op-ed pieces on the passenger pigeon reflect societal attitudes toward endangered species, extinction, and biodiversity conservation?

Method

This study uses framing theory to analyze social attitudes toward extinction and conservation as manifested in newspaper opinion pieces. As Entman (1993: 52) wrote: "To frame is to select some aspects of a perceived reality and make them more salient in a communicating text, in such a way as to promote a particular problem definition, causal interpretation, moral evaluation, and/or treatment recommendation for the item described." We believe media opinion pieces reflect the central social attitudes associated with passenger pigeons' extinction. By following a four-step framing process, our analysis reveals what kinds of information are made "noticeable, meaningful, or memorable" to audiences (Entman, 1993: 53), allowing us to investigate frames presented to the public and what attitudes the frames reflect to draw the public's attention and awareness to extinction of endangered species.

This study focused on all editorial or op-ed commentary in U.S. and Canadian newspapers from September 1, 2013, to October 21, 2016. Although studies have reported inconsistent evidence about differences and similarities between editorials and op-eds (Golan, 2013; Golan and Wanta, 2004; Day and Golan, 2005), we analyzed them as a single category because we focused on broad societal attitudes rather than on specific individual positions.

The start date marks one year before the 100th anniversary of Martha's death. Our search of the academic database NewsBank with a supplementary search of Google News turned up 88 pieces. We reviewed each item to determine whether it focused primarily on the passenger pigeon. We retained 39 opinion pieces for analysis. In addition to deductive coding, we analyzed the data inductively for emergent categories related to our research question. Then we conducted an axial coding to group the open codes into broader conceptual categories that are used in the findings section (Saldaña, 2015).

Findings

We looked at four ways in which pieces about passenger pigeon extinction were framed – problems defined, causes diagnosed, kinds of moral judgments made, and recommended remedies, if any.

Defining the problem

Initially, we examined how extinction was defined and whether it was referred to as an event specific to the passenger pigeon or as a broader environmental phenomenon posing a greater risk to multiple species. Most pieces defined extinction as a problem but in various contexts. Some simply used the word "extinction" loosely. Others (Turqman, 2014) more generally discussed consequences of extinction on ecosystems, the problem of "conflict," for example, the balance between abundance and extinction of resources, the view of passenger pigeons as pests, and conflicts between pigeons and farmers. While this aspect was not thoroughly explained, the authors clarified that because passenger pigeons traveled in huge flocks, they wreaked havoc on crops and trees where they landed. Passing flocks stripped seeds off crops and deposited enormous amounts of dung. Historical records indicate that entire trees collapsed under the weight of large flocks. Most connections between the ensuing conflict with farmers and extinction were implicit in the analyzed pieces.

Extinction as a problem was not always specifically defined in relation to the passenger pigeon. A few pieces defined a global biodiversity crisis as the problem. Martha and passenger pigeons in general appeared multiple times to illustrate something larger, more ominous, and urgent: global extinction. A *New York Times* piece (Fitzpatrick, 2014) suggested that the bluefin tuna is "the passenger pigeon of the sea," given how the last remnants of a once-abundant marine species are being plundered. The monarch butterfly, too, was said to potentially follow Martha's path to extinction. Some pieces also said the planet is currently at the cusp of what scientists label the sixth mass extinction – but the first one in human history and the first one that humans directly contribute to. "The largely man-made crisis [of extinction] may have already raised the natural or 'background' extinction rate by a factor of 1,000. Iconic animals like tigers, sharks, gorillas, and elephants could follow Martha if more isn't done to protect them," as the Mother Nature Network put it (McLendon, 2014). Such parallels between the passenger pigeon and present-day extinction possibilities used other examples, too, while stressing that it took less than 50 years for what was likely North America's most abundant bird species to plunge from 3 to 5 billion in the early to mid-1800s to zero in 1914. "Extinction was no longer just about dodo birds or rare species; it could happen to a creature once so plentiful that it was both a major food source and a pest of biblical proportions" (Eblen, 2014). Another piece published online by *Smithsonian* (Kutner, 2014) stated that Martha's demise could be viewed as an opportunity to reflect upon our problem with sharing our planet with other creatures.

Diagnosing causes

Most pieces attributed passenger pigeon extinction to human activities. Some used a justifying tone, positing that the extinction resulted from social progress or asserting that people lacked awareness of the abundant species. Primary anthropogenic contributing factors included overhunting, commodification of the birds as a food source, and the misperception that the species was so abundant and easy to capture and kill that extinction seemed far-fetched. Authors used evocative terms such as "mass slaughter" and the "holocaust suffered by passenger pigeons" to indicate the consistent level of exploitation and killing that eventually destroyed the species. Technological advances were blamed for overhunting, specifically the development of sophisticated firearms.

Besides overhunting and commodification as a food source, other commonly cited factors were habitat encroachment, habitat fragmentation, forest clearance for agriculture, habitat loss due to climate change, absence of laws or conservation movements to protect species, invasive species, pollution, degradation of migratory bird habitats, greed, lack of government commitment, and narrowly focused approaches to conservation that either disproportionately favor charismatic species or focus on individual species instead of protecting the overall habitat. Although not a common theme, more than one piece mentioned that passenger pigeons did not breed well in captivity, something that could have factored into their extinction. Discussions about naturally occurring causes such as diseases that could have contributed to extinction were rare.

Large-scale commercial destruction of the pigeons began in the 1840s and peaked in the late 1860s; billions were killed every year in this period but suggestions for legal protection confronted firm opposition from those who believed they needed no protection, given their abundance (Sutton, 2016). As deforestation spread, they lost roosting sites and sources of food. Industrialization brought the telegraph into the picture, along with railroads and an increased cache of firearms (Widmer, 2014), combined with a paucity of hunting regulations. The pigeons were docile, making it fairly easy for both subsistence hunters and professional market hunters to trap, kill, and sell them. Millions of dead birds were ferried in trains from capture sites to New York and Philadelphia restaurants (Fischetti, 2014). The passenger pigeon's "cheap meat fed slavery and industrialization" (Yeardley, 2014).

Notably however, not everything was pinned on unrestrained and brutal human beings. According to Fischetti (2014), "the pigeon's biology didn't help it survive the onslaught." Because they thwarted predators largely through strength in numbers, the author wrote that when their numbers ultimately fell, they lacked other means of coping with natural and human predators. Another piece (GrrlScientist, 2014) cited a study showing the pigeon was not always as abundant as records might have indicated, but rather "experienced dramatic population fluctuations in response to variations in annual acorn production." Ultimately, this study said, extinction likely resulted from the effects of these natural population fluctuations and anthropogenic activities. Their natural behavior – conspicuous roosting and breeding behavior – it said, might have made them easy targets of predators and hunters.

Moral judgments

Frames typically encompass moral judgments about problems, causes, and/or solutions. One major moral idea in these opinion pieces was that people tended to believe the supply of the species was endless. Authors advocated greater awareness and more knowledge of both abundant and endangered species to protect today's natural heritage for future generations. This moral theme also faulted humans for holding pigeons in contempt, barely worthy of notice.

These pieces also discussed that humans have the capacity and intellect to protect endangered species and the intrinsic value of nature. Some writers asserted that more advanced technologies now allow humans to reflect on the past and hope for the future, especially to protect species from the brink of extinction. In this respect, de-extinction was framed as a morally debatable issue. Authors asked whether it was a good idea to bring back species that are gone. They cited the potential dangers of attempts to control and manage such aspects of the environment. These authors argued that past wrongs cannot be righted and projects centered around de-extinction could divert valuable funds from conservation projects. Another moral argument against de-extinction was the "Pandora's box" argument, the suggestion that human interference can let loose a host of unintended problems greater than those it was meant to solve.

Remedies

Pieces discussed both individual-level and societal-level remedies for extinction problems. At the societal level, they noted that conservation-related laws came too late for species like the passenger pigeon, and thus contemporary species and habitats need more regulations, protection, and restoration. Overall, authors suggested that endangered species must be saved to restore global ecosystems. To save species, humans need persistent and continued economic investment and commitment. Some pieces suggested that new media and social networks' power to disseminate information and provide an ever-expanding, accessible platform of knowledge (the *Know-o-sphere*), could spread knowledge of de-extinction and restoration ecology, saving more in the future.

We found a noticeable lack of recommendations for more funding for conservation agencies. When funding was referred to, it was usually in the context of noting that projects involving charismatic species often win disproportionately more money than those for lesser-known species. One piece mentioned how governments' anti-environment and pro-oil or pro-industry attitudes affect the amount of research funds for such projects. Laws, such as the Endangered Species Act, Migratory Bird Treaty Act, and Lacey Act, were cited as remedies to today's extinction crisis. No analyzed piece detailed how those laws might help safeguard endangered species.

Another remedy recommended abundance over preventing extinction – focusing on protecting habitats instead of preventing endangered species from going extinct. Some pieces said today's numerous, more imminent threats to endangered species call for strong leadership in government and policy-making.

De-extinction also appeared as a remedy. One piece said neither conventional genetics nor recent advances in paleogenetics – reassembly of recently extinct animals' genomes – were likely to make de-extinction possible. Results of this science are not precise, with too many unknown variables. Furthermore, it is difficult to predict everything about an organism simply from its genome sequence. For example, young birds learn significant survival skills from adult birds, and a band-tailed pigeon's way of life differs greatly from that of a passenger pigeon. Thus even if – by a long stretch of imagination – a genetic replica of the passenger pigeon could be created, our best hope would be a bird that is partly band-tailed pigeon and partly passenger pigeon – a Band-Aid pigeon, as one piece suggests. Another piece said species such as the passenger pigeon and woolly mammoth should not be recreated merely because people might like to see them or to prove it can be done. Only one piece cited de-extinction as a possible remedy against extinction itself by mentioning initiatives like the Great Passenger Pigeon Comeback project and the idea of Revive and Restore (Louk, 2014). Another floated the idea that public oversight might remedy the risks posed by de-extinction projects.

Some pieces urged more public involvement in knowledge-raising and educational initiatives to help them adapt to a sustainable future.

Discussion and conclusions

Our examination of editorial pieces revealed several patterns of framing consistent with Entman's (1993) definition of how issue-oriented frames function compared to more abstract frames. They pinned responsibility for extinction almost exclusively on human activity without the broader range of causes found in the literature (Fuller, 2014), some of which narrowed the cause to Indgenous tribes or colonists. References to other causes occurred only twice, once to disease and once to the role of genetics in shaping the pigeon population. Specific human activities leading to extinction were overhunting, human consumption, and commodification of nature. When extinction was discussed, it was usually in terms of charismatic fauna such as monarch butterflies rather than ecosystems.

Moral judgments emerged strongly and consistently, with extinction consistently treated as caused exclusively or almost exclusively by human activity. This may reflect societal attitudes toward species conservation. This anthropogenic cause was consistently labeled as an unambiguous moral failure that should not be repeated. People are bad, the pieces asserted, and people have the capacity and intellect to be responsible for caring about nature; only carelessness or lack of foresight stands in the way. Slaughter of the pigeons was treated as relentless and merciless, an avian holocaust (Jacobs, 2016). It was also presented as a metaphor for the dangers of inattention to species survival and a warning to consider costs and consequences of human actions on fauna before it becomes too late.

However, these moral judgments did not lead to deeper discussions of policy remedies. Few policy recommendations emerged; those that did tended to refer to the continuing importance of relying on current legal structures such as the Endangered Species Act to prevent future extinctions. Unexplained were such specifics as more funding for agencies, the role of past environmental actions in future extinctions, the Lacey Act, the Migratory Bird Treaty Act, the

Know-o-sphere, technologically oriented solutions, or the role of strong leadership (Revkin, 2014; Fitzpatrick, 2014). The possibility of genetic technology to bring back extinct species was mentioned, but those pieces were ambivalent about chances for success.

Our analysis of the pieces, all of them issue-specific, found Entman's four components of frames emerging consistently. Each piece included at least one example of problem definition, cause attribution, moral evaluation, or remedy. Frequently there were more than one, and occasionally all four. Framing can describe this. However, the consistency of the framing pattern is better explained by framing's tendency to draw on an existing stock of broader cultural meanings. Pieces about the centennial of the passenger pigeon's extinction related to deeper themes, such as the special status of nature and the wild environment, the finality of extinction, and the need to preserve wildlife (McShea, 2014; Smith, 2014). Furthermore, this pattern occurred across a broad-enough spectrum of media to suggest that connections between frames and the culture occur regardless of the intent of individual media or journalists. That is, connections take place at an unconscious level and separately from the process that ensures journalistic accuracy and objectivity. Journalists appear unaware of this process that leaves them vulnerable to media manipulators and has a measurable effect on public opinion about the topic being framed.

These pieces consistently framed the extinction of the passenger pigeon – and by extension, threats to current and future endangered species – as a moral failure by humans. This failure was attributed to the utilitarian function of the species in a time when humans didn't know best. However, the present provides technological know-how to save endangered species, not only because of their utilitarian value (such as coral reefs for tourism) and eco-systemic value (such as bees for pollination), but because of their intrinsic value (the majestic nature of bald eagles). This reflects a shift in societal attitudes toward animals, particularly charismatic and iconic species. However, these pieces overlook an important component: an explanation of why conservation of endangered species is important for our own survival. This could ultimately prove the strongest motivator for a systemic commitment to conservation.

References

Anderson, A. (1997). *Media, Culture and the Environment*. New Brunswick, NJ: Rutgers University Press.

Audubon. (2017). *The Last Carolina Parakeet*. Available at http://johnjames.audubon.org/last-carolina-parakeet

Batt, S. (2009). Human attitudes towards animals in relation to species similarity to humans: A multivariate approach. *Bioscience Horizons*, 2: 180–190.

Beans, B. E. (1997). *Eagle's Plume: The Struggle to Preserve the Life and Haunts of America's Bald Eagle*. Lincoln, NE: University of Nebraska Press.

Ceballos, G., Ehrlich, P. R., Barnosky, A. D., García, A., Pringle, R. M., and Palmer, T. M. (2015). Accelerated modern human – induced species losses: Entering the sixth mass extinction. *Science Advances*, *1*: e1400253.

Ciofalo, A., and Traverso, K. (1994). Does the op-ed page have a chance to become a public forum? *Newspaper Research Journal*, *15*: 51–63.

Day, A., and Golan, G. (2005). Source and content diversity in op-ed pages: Assessing editorial strategies in the New York Times and the Washington Post. *Journalism Studies*, 6: 61–71.

De Pinho, J. R., Grilo, C., Boone, R. B., Galvin, K. A., and Snodgrass, J. G. (2014). Influence of aesthetic appreciation of wildlife species on attitudes towards their conservation in Kenyan agropastoralist communities. *PloS one*, 9: e88842.

Douglas, L. R., and Verissimo, D. (2013). Flagships or battleships: Deconstructing the relationship between social conflict and conservation flagship species. *Environment and Society*, *4*: 98–116.

Eblen, T. (2014). The anniversary of the last passenger pigeon, a bird that once filled Kentucky's skies. *Lexington Herald Leader*. Available at www.kentucky.com/news/local/news-columns-blogs/tom-eblen/article44505411.html

Entman, R. M. (1993). Framing: Toward clarification of a fractured paradigm. *Journal of Communication*, *43*: 51–58.

Feinberg, M., and Willer, R. (2013). The moral roots of environmental attitudes. *Psychological Science*, *24*(1): 56–62.

Fischetti, M. (2014). Passenger pigeons went extinct 100 years ago. *Scientific American*. Available at https://blogs.scientificamerican.com/observations/passenger-pigeons-went-extinct-100-years-ago/

Fishbein, M., and Ajzen, I. (1975). *Belief, Attitude, Intention and Behavior: An Introduction to Theory and Research*. Reading, MA: Addison-Wesley.

Fitzpatrick, J. W. (2014). Saving our birds. *New York Times*. Available at www.nytimes.com/2014/08/31/opinion/sunday/saving-our-birds.html

Freedman, E. (2011). Extinction is forever. *Earth Island Journal*. Available at www.earthisland.org/journal/index.php/eij/article/extinction_is_forever

Fuller, E. (2014). *The Passenger Pigeon*. Princeton, NJ: Princeton University Press.

George, K. A., Slagle, K. M., Wilson, R. S., Moeller, S. J., and Bruskotter, J. T. (2016). Changes in attitudes toward animals in the United States from 1978 to 2014. *Biological Conservation*, *201*: 237–242.

Golan, G. J. (2013). The gates of op-ed diplomacy: Newspaper framing the 2011 Egyptian revolution. *International Communication Gazette*, 1748048513482264.

Golan, G., and Wanta, W. (2004). Guest columns add diversity to NY Times' op-ed pages. *Newspaper Research Journal*, *25*(2): 70–82.

Greenberg, J. (2014). *A Feathered River Across the Sky: The Passenger Pigeon's Flight to Extinction*. New York: Bloomsbury Publishing USA.

GrrlScientist. (2014). Passenger pigeon extinction: It's complicated. *The Guardian*. Available at www.theguardian.com/science/grrlscientist/2014/jun/16/passenger-pigeon-extinction-population-fluctuations-human-exploitation

Jacobs, M. (2016). Always in season: Ubiquitous pigeons are newcomers here. *Grand Forks Herald*. Available at www.grandforksherald.com/outdoors/wildlife/3957581-always-season-ubiquitous-pigeons-are-newcomers-here

Kalof, L., and Fitzgerald, A. (2003). Reading the trophy: Exploring the display of dead animals in hunting magazines. *Visual Studies*, *18*: 112–122.

Kellert, S. R., and Westervelt, M. O. (1984). Children's attitudes, knowledge, and behaviors towards animals. *Children's Environments Quarterly*, *1*: 8–11.

Kerlinger, P., Burger, J., Cordell, H. K., Decker, D. J., Cole, D. N., Landres, P., Smith, E. N., Brett, J., Larson, R., and O'Shead, T. (eds.). (2013). *Wildlife and Recreationists: Coexistence Through Management and Research*. Washington, DC: Island Press.

Knight, S., Vru, A., Brandon, D., and Bard, K. (2010). Human rights, animal wrongs? Exploring attitudes toward animal use and possibilities for change. *Society and Animals*, *18*: 251–272.

Knight, S., Vru, A., Cherryman, J., and Nunkoosing, K. (2004). Attitudes towards animal use and belief in animal mind. *Anthrozoös*, *17*: 43–62.

Kutner, M. (2014). The most extensive report ever on American birds says there's cause for concern. *Smithsonian*. Available at www.smithsonianmag.com/smithsonian-institution/most-extensive-report-ever-published-american-birds-says-theres-cause-concern-180952644/

Leader-Williams, N., and Dublin, H. T. (2000). *Charismatic Megafauna as Flagship Species*. Cambridge, UK: Cambridge University Press.

Louk, M. S. (2014). Op-ed: De-extinction project aims to bring extinct species back to life. *Digital Journal*. Available at www.digitaljournal.com/science/op-ed-de-extinction- project-aims-to-bring-extinct-species-back-to-life/article/387181

McLendon, R. (2014). 100 years later, the passenger pigeon still haunts us. *Mother Nature Network*. Available at www.mnn.com/earth-matters/animals/blogs/100-years-later- the-passenger-pigeon-still-haunts-us

McShea, P. (2014). The great extermination: Passenger pigeons once filled the skies, but we killed them all. *Pittsburgh Post-Gazette*. Available at www.post- gazette.com/opinion/2014/06/29/The-great-extermination-Passenger-pigeons-once-filled- the-skies-but-we-killed-them-all/stories/201406290023

Mitman, T., Nikolaev, A., and Porpora, D. (2012). The critical moral voice on American newspaper opinion pages. *Communication, Culture and Critique*, 5: 392–408.

Parks, P., & Takahashi, B. (2016). From Apes to Whistleblowers: How Scientists Inform, Defend, and Excite in Newspaper Op-Eds. *Science Communication*, 38(3): 275–302.

Petersen, S. (1999). Congress and charismatic megafauna: A legislative history of the Endangered Species Act. *Environmental Law*, *29*: 463–491.

Raymond, L., and Schneider, L. U. (2014). Personal moral norms and attitudes toward endangered species policies on private land. *Conservation and Society*, *12*: 1.

Revkin, A. C. (2014). The role of social media in wiping out passenger pigeons, and conserving species now. *New York Times*. Available at http://dotearth.blogs.nytimes.com/2014/09/01/the-role-of-social-media-in-wiping-out- passenger-pigeons-and-conserving-species-now/

Saldaña, J. (2015). *The Coding Manual for Qualitative Researchers*. Thousand Oaks, CA: Sage.

Sax, J. L. (2001). The new age of environmental restoration. *Washburn Law Journal*, *41*: 1–13.

Schlegel, J., and Rupf, R. (2010). Attitudes towards potential animal flagship species in nature conservation: a survey among students of different educational institutions. *Journal for Nature Conservation*, *18*: 278–290.

Schulz, J. H., Otis, D. L., and Temple, S. A. (2014). 100th anniversary of the passenger pigeon extinction: Lessons for a complex and uncertain future. *Wildlife Society Bulletin*, 38: 445–450.

Smith, P. A. (2014). 100th anniversary of passenger pigeon's extinction reveals lessons. *Journal Sentinel*. Available at http://archive.jsonline.com/sports/outdoors/100th-anniversary-of- passenger-pigeons-extinction-reveals-lessons-b99264809z1–258399811.html

Sommer, B., and Maycroft, J. R. (2008). Influencing public policy: An analysis of published op- eds by academics. *Politics & Policy*, 36: 586–613.

Stein, B. A., Kutner, L. S., and Adams, J. S. (2000). *Precious Heritage: The Status of Biodiversity in the United States*. Oxford, UK: Oxford University Press.

Stork, N. E. (2009). Re-assessing current extinction rates. *Biodiversity and Conservation*, *19*: 357–371.

Sutton, K. (2016). The sad story of the passenger pigeon. *Three Rivers Edition*. Available at www.arkansasonline.com/news/2016/oct/16/sad-story-passenger- pigeon/?f=threerivers

Turqman, M. (2014). Remembering Martha, the last passenger pigeon. *National Geographic: Education Blog* [Online]. Available at https://blog.education.nationalgeo graphic.com/2014/09/11/remembering-martha- washington-the-last-passenger-pigeon/

Widmer, T. (2014). The Civil War's environmental impact. *New York Times*. Available at http://opinionator.blogs.nytimes.com/2014/11/15/the-civil-wars-environmental-impact/? _r=0

Yeardley, D. (2014). Lost birds. *CounterPunch*. Available at www.counterpunch.org/2014/10/10/lost-birds/

Young, H. S., McCauley, D. J., Galetti, M., and Dirzo, R. (2016). Patterns, causes, and consequences of anthropocene defaunation. *Annual Review of Ecology, Evolution, and Systematics*, *47*: 333–358.

7 Recovering the ecology of fear

Cascading effects of gray wolf predation and competition in a Great Lakes Basin forest

David G. Flagel

Introduction

We have been competing with wild predators to catch and consume ungulates such as deer for ages. Over time, humans have gained the upper hand in this conflict, becoming adept at not only securing game but eliminating our competitors (DelGiudice et al., 2009). Large carnivores have been reduced or removed from several ecosystems by habitat loss, eradication efforts, or other factors. Large carnivore eradication is not without consequence as these species may have substantial impacts on food web dynamics, community structure, and biodiversity through interactions with their prey and their own natural competitors (Estes et al., 2011; Ripple et al., 2014).

Trophic cascades are chain interactions that occur when a top predator regulates another consumer species in such a way that it ends up benefitting the resources of that consumer – a natural version of "the enemy of my enemy is my friend" (Hairston et al., 1960; Estes et al., 2011). Classic examples focus on interactions where a predator benefits producers by killing an herbivore. Trophic cascades also include chain interactions where larger predator A kills predator B, benefiting any species that B eats or harasses. Originally thought to be rare in terrestrial systems, trophic cascades are now documented in ecosystems across the globe (Estes et al., 2011; Ripple et al., 2014). Predators can produce trophic cascades through two mechanisms. A density-mediated trophic cascade occurs when predators benefit a resource by directly controlling the numbers of their prey or competitors. In contrast, a behaviorally-mediated cascade develops when predators induce fear responses in their prey/competitor, such as spatial avoidance and sacrificing feeding time for vigilance, and thus reducing their foraging activity (Lima and Dill, 1990; Polis and Strong, 1996; Werner and Peacor, 2003).

Removing large carnivores removes their top-down influences. The widespread extirpation of the gray wolf (*Canis lupus*) from much of the Great Lakes Basin serves as a notable example (Ripple et al., 2014). Wolves were once fairly ubiquitous throughout the Basin as a top predator of ungulates, including white-tailed deer (*Odocoileus virginianus*) and moose (*Alces alces*) (DelGiudice et al., 2009). By 1960, they were mostly exterminated through habitat change, poisoning, and bounty hunting (Thiel, 1993). In their absence, white-tailed deer numbers increased substantially due to reduced large carnivore predation, favorable hunting regulations, habitat changes, and warming winters. High deer populations in a landscape without their top predator are now significant drivers of environmental change in these and many other forests in eastern North America, inhibiting sapling growth and eliminating sensitive wildflower species from the understory (Alverson et al., 1988; Côte et al., 2004).

Wolves also kill and exclude coyotes (*Canis latrans*), their primary midsized competitors for fawns and scavengers of wolf kills. This species prefers open habitats such as grasslands and arid environments, where it feeds on ungulate fawns, rabbits, and larger rodents. However, coinciding with the removal of wolves by humans and the opening of habitat for settlement and agriculture, the coyote expanded its range eastward from the Great Plains into the forests of the Great Lakes Basin (Gompper, 2002; Ripple et al., 2013). Since then, coyotes have likely had substantial negative impacts on the density and behavior of their prey – particularly rabbits such as snowshoe hares (*Lepus americanus*) – and smaller competing carnivores, especially foxes (Ripple et al., 2013). Negative impacts on these small carnivores also likely benefit their primary prey, deer mice (*Peromyscus* spp.), which may affect Lyme disease risk given that mice are major disease reservoirs (Levi et al., 2012).

Mitigating and/or reversing the adverse impacts of high deer and coyote populations is a major conservation and economic priority (Côte et al., 2004; Ripple et al., 2013). The most widely used management tool for dealing with deer impacts in the United States is recreational hunting, but populations remain high and negative impacts on forest regeneration and understory diversity continue (Côte et al., 2004). Hunting is also important for coyote control, but this species is resilient against reductions and quickly replenishes its numbers (Knowlton et al., 1999; Gompper, 2002). Alternative methods such as sharpshooting, frightening devices, and reproductive interference have been proposed but are labor-intensive, expensive, difficult to scale up, or simply ineffective (Knowlton et al., 1999; Côte et al., 2004).

Wolf recovery has been proposed as an additional alternative on the basis of research from key national parks (Licht et al., 2010; Ritchie et al., 2012; Ripple et al., 2013). Wolf recovery is thought to have produced a measurable trophic cascade in Yellowstone and Banff national parks through combinations of lethal and behavioral effects on elk (*Cervus canadensis*) which have benefitted aspen growth (*Populus* spp.) (Hebblewhite and Smith, 2010; Ripple and Beschta, 2012). Research at Isle Royale National Park suggests wolves indirectly affect balsam fir (*Abies balsamea*) via declines in moose populations (McLaren and Peterson, 1994). Yellowstone wolves have also impacted the density, distribution, and behavior of coyotes, with limited evidence of benefits to voles and pronghorn fawns (*Antilocapra americana*) (Berger et al., 2008; Miller et al., 2012; Ripple et al., 2013). Isle Royale wolves, shortly after colonization, also eliminated coyotes from the island (Krefting, 1969). Few studies, however, have tested whether any of these findings will transfer to other systems being recolonized by wolves outside of national parks.

Hypotheses

Although it is tempting to assume wolves will generate similar effects in other systems, trophic cascades are context-dependent interactions (Pace et al., 1999), varying with differences in ecosystem-specific characteristics such as species composition, regional behavior, and system structure (Pace et al., 1999; Schmitz, 2005). The mainland of the Great Lakes Basin presents a unique opportunity to evaluate the generality of wolf-generated trophic cascades. Wolves have naturally recolonized the northern forests of the Great Lakes states, which have high deer populations and prevalent coyotes (DelGiudice et al., 2009; Levi et al., 2012).

The aims of our research were to (1) determine if there is evidence for density- and/or behaviorally mediated effects of wolves on deer herbivory; (2) evaluate if these wolf–deer interactions affect plant growth and diversity; (3) document whether wolf presence simultaneously affects coyote and fox distribution; and (4) assess whether these wolf–coyote interactions cascade to the major prey of coyotes (hares) and foxes (deer mice). Figure 7.1 diagrams a simplified version of these food web interactions.

Specifically, we hypothesized that in areas used heavily by wolves:

1. Deer would visit plots less frequently and for shorter time periods.
2. Coyotes would visit less frequently.
3. Foxes would visit more frequently.
4. Deer mice would be less prevalent on plots.
5. Deer would sacrifice foraging time for other activities, such as vigilance.
6. Deer would browse a lower proportion of saplings.
7. Snowshoe hares would browse a higher number of saplings.
8. Maple saplings, as a species that deer prefer to eat, would exhibit greater growth and recruitment of individuals to a taller size class.
9. There would be higher wildflower diversity.
10. The impacts of white-tailed deer would be measurably reduced based on experimental tests of deer impact strength.

Method

Our study took place at the University of Notre Dame Environmental Research Center (UNDERC; ~ 46°13′N, 89°31′W). This ~ 3,200-hectare (7,900-acre) property along the Wisconsin–Michigan border consists of a scattered array of northern mesic forest patches, wetlands, and bogs. A mixture of sugar maple (*Acer saccharum*) and red maple (*Acer rubrum*), which deer readily consume, makes up the main deciduous component. Balsam fir, which white-tailed deer largely avoid consuming, makes up the primary coniferous component. After several years of high deer numbers, Canada mayflower (*Maianthemum canadense*) and Canadian bunchberry (*Cornus canadensis*) dominate wildflower composition. White-tailed deer are the

Figure 7.1 Simplified interaction diagram for cascading effects of wolves at the University of Notre Dame Environmental Research Center.

only ungulate permanently living there. Population densities based on Wisconsin Department of Natural Resources deer management unit estimates are high enough to cause significant herbivory damage to the plant community (Alverson et al., 1988).

Gray wolves are the natural top predator of these deer (Mech and Karns, 1977). By 1956, wolves were likely extirpated from UNDERC (Arnold and Schofield, 1956), but recolonizing wolves established a territory that bisects the UNDERC property during or after 2002. Additional packs use areas surrounding the property, but these territories do not appear to overlap with UNDERC or its pack. Wolves often establish "buffer zones" between their territories that they avoid entering to prevent conflict (Mech, 1977). These areas represent low-/no-wolf-use areas and cover a substantial portion of UNDERC. Coyotes also likely moved into this area sometime during the late 1800s or early 1900s (Thiel, 1993). Two fox species, red (*Vulpes vulpes*) and gray (*Urocyon cinereoargenteus*), are also present. Snowshoe hares are the most common lagomorph species, while deer mice are the dominant small rodents.

We employed variable wolf use as our treatment effect. Areas of high-wolf-use are generally located within the wolf territory, as determined through collar telemetry for tracking and surveys of wolf sign. We selected ten forest patches as test sites in 2009 and an additional two in 2011. These sites were distributed evenly among high- and low-wolf-use areas, with an adjacent pair of control (open to deer) and experimental (exclosure; enclosed by fencing to keep out deer) plots established at each one (10 meters × 10 meters, 33 feet × 33 feet).

We monitored deer activity on high- and low-wolf-use control plots using motion-activated, infrared-flash camera traps and browse damage surveys. Cameras were deployed for three weeks on ten sites, spanning June through August 2012. We used images to quantify deer visits as a measure of density, the duration of visits, and the proportion of time spent foraging (head down). We monitored deer impacts on all plots annually in August for maple sapling height (2009–2012) and wildflower diversity (2010–2012). Surveys recorded all sapling heights (individuals 25–240 centimeters [9.8–94.5 inches]), deer browse damage on saplings, and wildflower species richness (20 1-square-meter [10.7-square-feet] subplots per plot). The percentage of maple saplings browsed by deer was used as an index of deer foraging pressure. Our exclosure design meanwhile allowed us to calculate and compare the strength of deer impacts between high- and low-wolf-use areas using the log-response ratio. This ratio is the log (X_e/X_c), where X_e represents the exclosure plot average and X_c represents the control plot average (see Flagel et al., 2016 for detailed wolf–deer methods).

To determine effects on predator distribution, we used mapped distributions of coyote and fox signs from 2008 and 2010–2014, comparing the frequency of points in high- versus low-wolf-use areas. All accessible gravel roads (> 37 kilometers [23 miles]; usually once weekly) were surveyed for carnivore scats from May–October annually except 2009. Other opportunistic evidence of carnivore use, such as sightings and tracks, were pooled into the sign dataset. We also noted hare browse damage on saplings at the deer exclosure sites as an index of hare density. Mark-recapture surveys in late summer (late July–early September, 2011–2013) estimated deer mouse abundance, combining data from plot grids of live traps (16

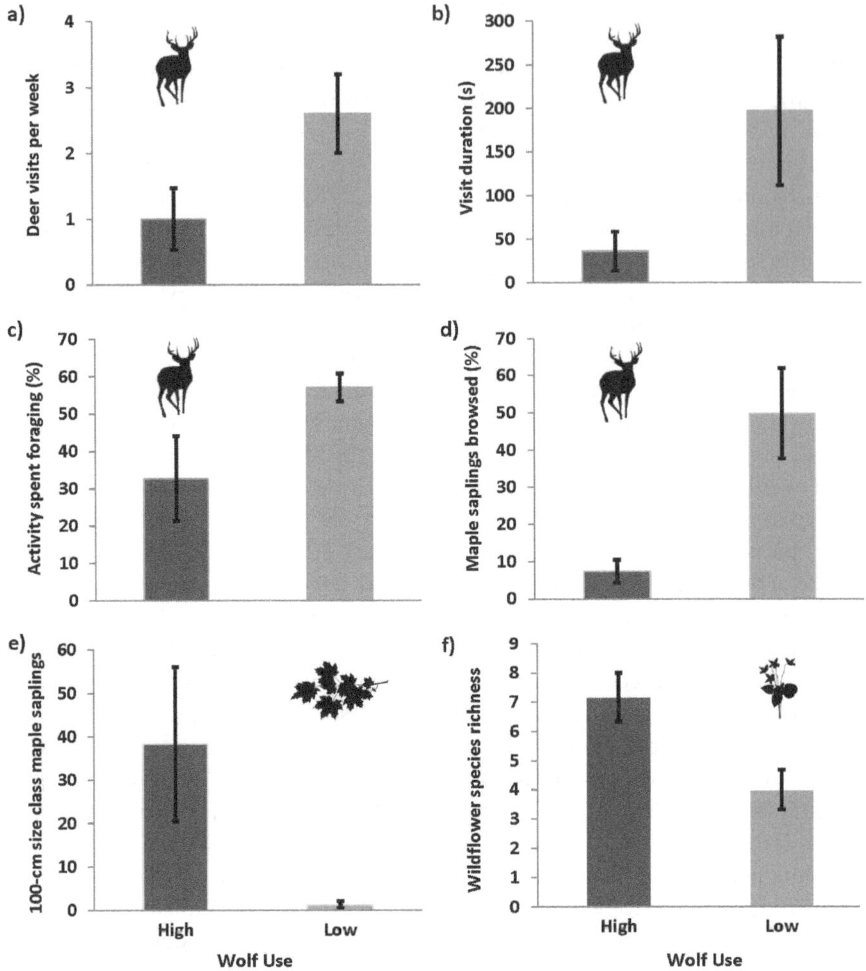

Figure 7.2 Comparisons of high-wolf-use and low-wolf-use control plot values for white-tailed deer: (a) visit frequency, (b) visit duration, (c) foraging activity, and (d) maple browse intensity, as well as (e) 100-centimeter size class maple sapling abundance and (f) wildflower species richness. Bars are means ±SE.

Adapted and modified from Flagel et al., 2016.

traps per site). Trapping was done for seven consecutive days over a two- to three-week period (see Flagel et al., 2017 for detailed wolf–coyote methods).

Findings

Based on *t*-tests of camera trap data, deer visited high-wolf-use plots substantially less frequently (1.0 versus 2.6 visits per week, $df = 8$, $P = 0.04$, Figure 7.2a).

These visits were also substantially shorter, averaging only 36 seconds in high-wolf-use areas compared to 197 seconds in low-wolf-use areas (*df* = 8, $P = 0.03$, Figure 7.2b). Deer spent only 33 percent of their time foraging in high-wolf-use areas, compared to 57 percent in low-wolf-use areas (*df* = 6, $P = 0.02$, Figure 7.2c), appearing to sacrifice this time for other activities such as vigilance.

We focused our wolf–deer analyses on maples (*Acer* spp.; > 90% of deer-preferred saplings) to avoid comparing species which deer consider differently as palatable and have differing inherent growth rates. Deer browsed nearly 50 percent of maple saplings on low-wolf-use control plots, but only about 7 percent of maple saplings on high-wolf-use control plots (*t*-test, *df* = 10, $P = 0.01$, Figure 7.2d). Average maple sapling height in high-wolf-use control plots was nearly 41 centimeters (16 inches) greater than in low-wolf-use plots, with low-wolf-use maple saplings hovering around 30 centimeters (12 inches) (*t*-test, *df* = 10, $P < 0.01$). As a measure of recruitment to the canopy, we also compared saplings that had been recruited to the 100-centimeter (39-inch) size class using a Mann-Whitney *U* test. These were almost exclusively found in high-wolf-use plots, with low-wolf-use plots averaging only ~1 per plot versus ~38 in high-wolf-use (*df* = 10, $P = 0.01$, Figure 7.2e). Meanwhile, high-wolf-use controls had about seven wildflower species per plot versus only four per plot for low-wolf-use controls (*t*-test, *df* = 10, $P = 0.01$, Figure 7.2f).

Based on repeated-measures ANOVAs of exclosure-control log-response ratios, deer had significant negative impacts on maple sapling growth in low-wolf-use areas (higher exclosure plot growth vs. control; $F_{3,12} = 4.08$, $P = 0.03$), while in high-wolf-use areas deer had no measurable impact on growth (growth equal; $F_{3,12} = 1.62$, $P = 0.24$). This suggests that negative impacts of deer on maple growth effectively disappeared in high-wolf-use areas (wolf use–year interaction, $F_{3,24} = 6.18$, $P < 0.01$). Deer impacts on wildflower species richness were also diminished in areas of high- compared to low-wolf-use (wolf-use factor, $F_{1,8} = 6.64$, $P = 0.03$).

Tests on coyote and fox signs were adjusted prior to analysis to account for there being more low-wolf-use road than high-wolf-use road. Based on a chi-square test of coyote sign distribution, coyotes used high-wolf-use areas 51 percent less often than low-wolf-use areas ($\chi^2 = 9.98$, *df* = *1*, $P < 0.01$, Figure 7.3a). Conversely, foxes used these areas almost exclusively (exact binomial probability, $P < 0.01$, Figure 7.3a). Total saplings browsed by hares were pooled by site (exclosure plot and adjacent control plot) as deer exclosure fences were made to be hare-accessible. Hare browse was three times more prevalent on high-wolf-use sites than low-wolf-use sites (6.7 vs. 2.0, t-test, *df* = 10, $P = 0.02$, Figure 7.3b), suggesting greater hare densities. Deer mouse abundance was significantly affected by wolf use, but the effect varied by year (repeated-measures ANOVA, wolf use–year interaction, $F_{2,20} = 8.05$, $P < 0.01$, Figure 7.3c). Deer mouse abundance was significantly less on high-wolf-use plots than low-wolf-use plots in 2011 (6.0 vs. 11.5, *t*-test, *df* = 10, $P = 0.04$) and 2013 (3.8 vs. 10.0, *t*-test, *df* = 10, $P = 0.03$), but not in 2012 (14.2 vs. 10.5, *t*-test, *df* = 10, $P = 0.22$).

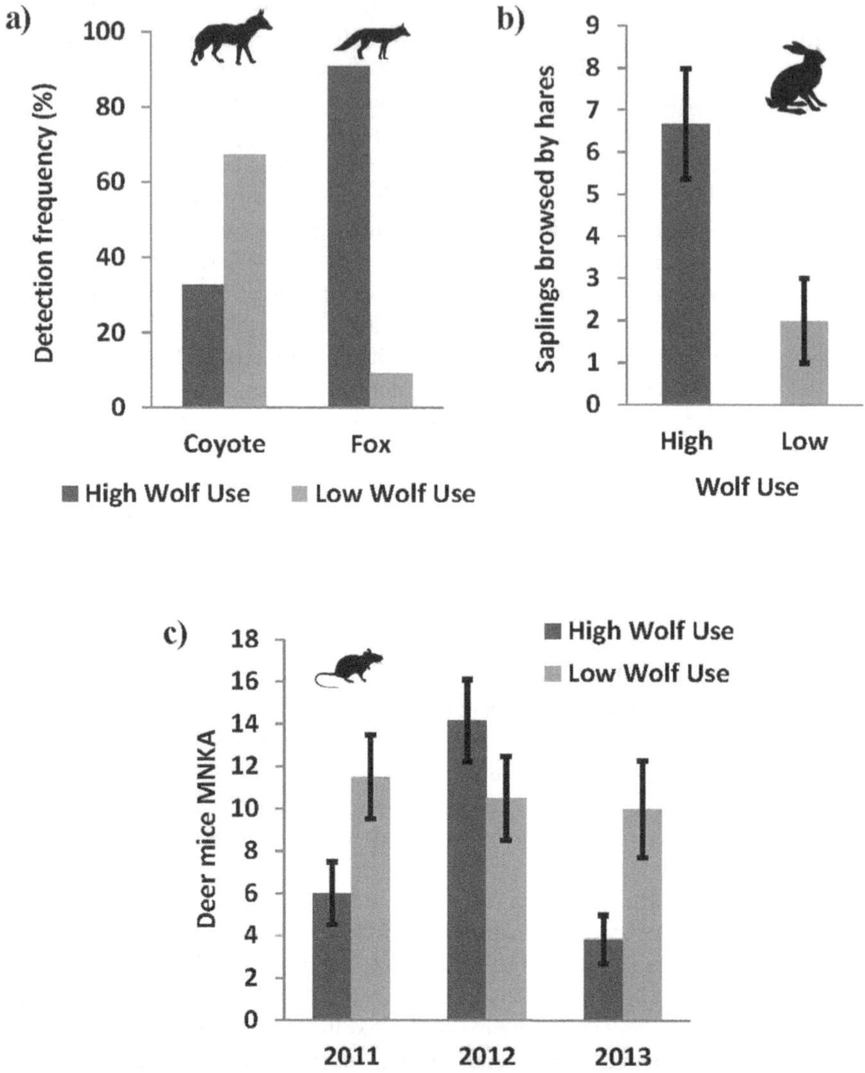

Figure 7.3 Comparisons of high-wolf-use and low-wolf-use values for (a) coyote and fox sign distribution, (b) exclosure site sapling browse by hares, and (c) exclosure site deer mice abundance. Bars are means ±SE.

Adapted and modified from Flagel et al., 2017.

Conclusion

This research is among the first to: (1) provide evidence for wolf–ungulate trophic cascades using manipulative tests; (2) document the behavioral mechanisms thought to drive the cascade; (3) test wolf–coyote–fox cascade hypotheses at the local scale; (4) document greater fox and hare use of high-wolf-use areas; and (5) observe reduced forest rodent abundances in high-wolf-use areas.

These results strongly suggest that gray wolf recovery may alter the density and behavior of white-tailed deer in a manner that indirectly benefits sapling growth and wildflower diversity as part of a trophic cascade. Deer density and behavior were significantly different in high-wolf-use areas than low-wolf-use areas. Deer density, visit length and foraging activity were lower by 62 percent, 82 percent, and 43 percent, respectively. These mechanisms for reduced herbivory appear to directly translate into different impacts on plants in high-wolf-use areas. By 2012, maple sapling browse was reduced by 85 percent, maple sapling height more than doubled, and recruitment of maples to the 100-centimeter size class was 24 times higher. Wildflower species richness also had nearly doubled. Based on comparisons using our manipulative exclosure experiment, it appears the negative impacts of deer on sapling growth and wildflower diversity normally seen in Great Lakes forests effectively disappeared in the high-wolf-use areas at UNDERC.

These results also strongly suggest that wolf recovery may generate trophic cascades benefitting smaller carnivores and midsized prey through changes in the density and distribution of coyotes. Coyotes used high-wolf-use areas half as much as low-wolf-use areas, while foxes used high-wolf-use areas almost exclusively. Hare browse, meanwhile, was three times greater in high-wolf-use areas in 2012, while deer mice were 48 percent and 62 percent less abundant in these areas in 2011 and 2013, respectively. These findings are consistent with the hypothesis that wolves generate trophic cascades via top-down impacts on coyotes.

Taken together, our research supports trophic cascade hypotheses involving wolves and other mammalian consumers in the Great Lakes Basin (Rooney and Anderson, 2009; Levi and Wilmers, 2012; Callan et al., 2013). It also supports, to a degree, the generality of Yellowstone and Banff findings. However, some of these observed wolf–deer cascade impacts from UNDERC appear stronger than comparable measures from wolf–elk cascade studies in the Rockies (see Flagel et al., 2016 for comparisons). This may be due to differences in the relative importance of behavioral effects, as reviews suggest behaviorally-mediated cascades should be stronger than density-mediated trophic cascades (Schmitz et al., 2004; Preisser et al., 2005). Great Lakes wolf–deer cascades are predicted to be primarily behaviorally mediated (Rooney and Anderson, 2009), which our documentation of strong behavioral effects supports, while evidence implies behavioral effects are weak at Yellowstone (Kauffman et al., 2010).

Our wildflower diversity findings also align with cascade research in northern white cedar wetlands in the Great Lakes region (Callan et al., 2013), while our carnivore distribution results are consistent with regional- and continental-scale population trends for foxes in areas with and without wolves (Levi and Wilmers, 2012; Newsome and Ripple, 2015). However, we cannot deduce from our findings whether wolf–coyote cascades are driven primarily by kills or behavioral effects. Wolves are known to kill coyotes and generate avoidance behaviors in the Great Lakes region (Krefting, 1969; Berg and Chesness, 1978), so both may be occurring. Little work has been done on the impact of wolf–coyote trophic cascades on plants. However, our results suggest total browse damage by hare in high-wolf-use areas does not eclipse that of the deer, which already have negligible effects and yet are the superior herbivore competitor to hares (Bookhout, 1965). Therefore, we do not predict that wolf–coyote cascades will significantly affect plants through hares. Impacts through rodents seem more likely because small mammals are an important factor in eastern forest seedling recruitment (Schnurr et al., 2004).

Granted, there are other possible explanations for our observations, such as potential differing habitat or abiotic conditions between high- and low-wolf-use areas that could produce similar results (Hebblewhite and Smith, 2010; Kauffman et al., 2010). To address this concern, we reran our analysis on sapling species that deer do not prefer to consume (such as balsam fir, *Prunus serotina*, *Picea* spp., etc.). Changes in deer herbivory should not strongly affect non-preferred species (White, 2012) but should react to changes in abiotics. We found no significant differences in non-preferred species' growth, unlike with maples. Maple sapling density and relative abundance, visibility, escape impediments, and terrain were also each relatively equivalent between high- and low-wolf-use area sites during the study (Flagel et al., 2016). As such, there is little to support alternatives to the trophic cascade hypotheses. As for why deer mice were affected in 2011 and 2013, but not 2012, we suspect they shifted from top-down regulation (fox predation) to bottom-up regulation (where factors such as space become limiting) during that year following beneficial weather and a heavy maple seed crop (Ostfeld and Keesing, 2000; Falls et al., 2007).

Our work continues at UNDERC, with plant surveys on a three-year cycle for the remaining life span of the deer exclosure network (up to 30 years). Results from the 2015 survey suggest that wolves continue to benefit maple sapling growth, with more individuals reaching the 100-centimeter (39-inch) size class (recruitment measure) and several growing beyond 240 centimeters (94 inches) (maximum browse reach of deer) since 2012. Wildflower diversity remains low overall, but new species such as nodding trillium (*Trillium cernuum*) have begun to appear here and there in exclosure plots and high-wolf-use areas. Browse patterns have also remained largely consistent with cascade hypotheses. Deer browse remains substantially reduced in high-wolf-use areas. Total hare browse has become particularly more prevalent in high-wolf-use areas since 2012, perhaps suggesting increased numbers. Based on continuing sign surveys, wolf use patterns from 2014–2016 were consistent with 2008–2013. The distribution of coyote use in 2015–2016 was slightly more biased towards low-wolf-use areas than the 2008–2014 average, which coincided with an increase in the amount of high-wolf-use fox sign. Subsequent deer mice surveys have yet to be done on the exclosure sites.

While human hunting is often the primary method for controlling ungulate and coyote impacts, high populations of these species continue to have negative effects on forest ecosystems (Côte et al., 2004; Ripple et al., 2013). Coyotes are quite resilient to human hunting (Knowlton et al., 1999; Gompper, 2002), and our usual deer hunting methods do not fully generate the perpetual fear responses necessary for reducing deer impacts (Cromsigt et al., 2016). Wolf recovery may be a useful additional measure for dealing with negative impacts of high ungulate and mesopredator populations (Licht et al., 2010; Ritchie et al., 2012; Ripple et al., 2013). However, we need further research so that we can predict when and where recovery will lead to cascading effects, and what those effects will be. We particularly need more study of trophic cascades in human-modified landscapes, outside national parks and other protected landscapes. Human impacts on the ability of wolves and other large carnivores to generate trophic cascades is also a key gap in our knowledge. For example, ungulates and coyotes may use areas of human activity to avoid wolves (Ripple et al., 2013; Kuijper et al., 2016). Breeder loss through human hunting of wolves could also disrupt pack social structure, potentially leading to pack dissolution and changes in spatial arrangement (Brainerd et al., 2008).

Large carnivore management is costly and controversial, and its long-term success or failure in human-occupied landscapes is inherently intertwined with human tolerance. Any recovery efforts should work with the public to ensure that stakeholders are heard and represented so that human–wildlife conflicts are minimized. This may reduce human impacts on the recovering predator population (Carter and Linnell, 2016). In doing so, we may find success and balance for the predator, the people, and the ecosystem.

References

Alverson, W. S., Waller, D. M., and Solheim, S. L. (1988). Forests too deer: Edge effects in northern Wisconsin. *Conservation Biology*, *2*(4): 348–358.

Arnold, D., and Schofield, R. (1956). *Status of Michigan Timber Wolves, 1954–56*. Michigan Department of Conservation, Game Division, Report No. 2097, mimeo, p. 2.

Berg, W. E., and Chesness, R. A. (1978). Ecology of coyotes in northern Minnesota. In: Bekoff, M. (Ed.), *Coyotes: Biology, Behavior, and Management*. New York: Academic Press, 158–172.

Berger, K. M., Gese, E. M., and Berger, J. (2008). Indirect effects and traditional trophic cascades: A test involving wolves, coyotes, and pronghorn. *Ecology*, *89*(3): 818–828.

Bookhout, T. A. (1965). *The Snowshoe Hare in Upper Michigan: Its Biology and Feeding Coactions with White-Tailed Deer*. Michigan Department of Conservation Research Development Report 38, p. 198.

Brainerd, S. M., Andrén, H., Bangs, E. E., Bradley, E. H., Fontaine, J. A., Hall, W., Iliopoulos, Y., Jimenez, M. D., Jozwiak, E. A., Liberg, O., Mack, C. M., Meier, T. J., Niemeyer, C. C., Pedersen, H. C., Sand, H., Schultz, R. N., Smith, D. W., Wabakken, P. and Wydeven, A. P. (2008). The effects of breeder loss on wolves. *Journal of Wildlife Management*, *72*(1): 89–98.

Callan, R., Nibbelink, N. P., Rooney, T. P., Wiedenhoeft, J. E., and Wydeven, A. P. (2013). Recolonizing wolves trigger a trophic cascade in Wisconsin (USA). *Journal of Ecology*, *101*(4): 837–845.

Carter, N. H., and Linnell, J. D. C. (2016). Co-adaptation is key to coexisting with large carnivores. *Trends in Ecology and Evolution*, *31*(8): 575–578.

Côte, S. D., Rooney, T. P., Tremblay, J. P., Dussault, C., and Waller, D. M. (2004). Ecological impacts of deer overabundance. *Annual Review of Ecology, Evolution, and Systematics*, 35: 113–147.

Cromsigt, J. P. G. M., Kuijper, D. P. J., Adam, M., Beschta, R. L., Churski, M., Eycott, A., Churski, M., Eycott, A., Kerley, G. I. H., Mysterud, A., Schmidt, K., and West, K. (2016). Hunting for fear: Innovating management of human-wildlife conflicts. *Journal of Applied Ecology*, *50*(3): 544–549.

DelGiudice, G. D., McCaffery, K. R., Beyer, D. E., and Nelson, M. E. (2009). Prey of wolves in the Great Lakes region. In: Wydeven, A. P., Van Deelen, T. R., and Heske, E. J. (Eds.), *Recovery of Gray Wolves in the Great Lakes Region of the United States: An Endangered Species Success Story*. New York, NY: Springer Science, 155–174.

Estes, J. A., Terborgh, J., Brashares, J. S., Power, M. E., Berger, J., Bond, W. J., Carpenter, S. R., Essington, E. E., Holt, R. D., Jackson, J. B. C., Marquis, R. J., Oksanen, L., Oksanen, T., Paine, R. T., Pikitch, E. K., Ripple, W. J., Sandin, S. A., Scheffer, M., Schoener, T. W., Shurin, J. B. Sinclair, A. R. E., Soulé, M. E., Virtanen, R., and Wardle, D. A. (2011). Trophic downgrading of planet Earth. *Science*, *333*: 301–306.

Falls, J. B., Falls, E. A., and Fryxell, J. M. (2007). Fluctuations of deer mice in Ontario in relation to seed crops. *Ecological Monographs*, *77*(1): 19–32.

Flagel, D. G., Belovsky, G. E., and Beyer Jr., D. E. (2016). Natural and experimental tests of trophic cascades: Gray wolves and white-tailed deer in a Great Lakes forest. *Oecologia*, *180*: 1183–1194.

Flagel, D. G., Belovsky, G. E., Cramer, M. J., Beyer Jr., D. E., and Robertson, K. E. (2017). Fear and loathing in a Great Lakes forest: Cascading effects of competition between wolves and coyotes. *Journal of Mammalogy*, 98(1): 77–84.

Gompper, M. E. (2002). Top carnivores in the suburbs? Ecological and conservation issues raised by colonization of north-eastern North America by coyotes. *BioScience*, *52*: 185–190.

Hairston, N. G., Smith, F. E., and Slobodkin, L. B. (1960). Community structure, population control, and competition. *The American Naturalist*, *94*: 421–425.

Hebblewhite, M., and Smith, D. W. (2010). Wolf community ecology: Ecosystem effects of recovering wolves in Banff and Yellowstone National Parks. In: Musiani, M., Boitani, L., and Paquet, P. C. (Eds.), *The World of Wolves: New Perspectives on Ecology, Behaviour, and Management*. Calgary, AB: University of Calgary Press, 69–120.

Kauffman, M. J., Brodie, J. F., and Jules, E. S. (2010). Are wolves saving Yellowstone's aspen? A landscape-level test of a behaviorally mediated trophic cascade. *Ecology*, *91*(9): 2742–2755.

Knowlton, F. F., Gese, E. M., and Jaeger, M. M. (1999). Coyote depredation control: An interface between biology and management. *Journal of Range Management*, *52*: 398–412.

Krefting, L. W. (1969). The rise and fall of the coyote on Isle Royale. *Naturalist*, *20*: 24–31.

Kuijper, D. P. J., Sahlén, E., Elmhagen, B., Chamaillé-Jammes, S., Sand, H., Lone, K., and Cromsigt, J. P. G. M. (2016). Paws without claws? Ecological effects of large carnivores in anthropogenic landscapes. *Proceedings of the Royal Society* B, *283*: 20161625.

Levi, T., Kilpatrick, A. M., Mangel, M., and Wilmers, C. C. (2012). Deer, predators, and the emergence of Lyme disease. *Proceedings of the National Academy of Sciences of the United States of America*, *109*: 1094210947.

Levi, T., and Wilmers, C. C. (2012). Wolves-coyotes-foxes: A cascade among carnivores. *Ecology*, 93(4): 921–929.

Licht, D. S., Millspaugh, J. J., Kunkel, K. E., Kochanny, C. O., and Peterson, R. O. (2010). Using small populations of wolves for ecosystem restoration and stewardship. *BioScience*, 60(2): 147–153.

Lima, S. L., and Dill, L. M. (1990). Behavioral decisions made under the risk of predation: A review and prospectus. *Canadian Journal of Zoology*, 68: 619–640.

McLaren, B. E., and Peterson, R. O. (1994). Wolves, moose, and tree rings on Isle Royale. *Science*, *266*: 1555–1558.

Mech, L. D. (1977). Wolf-pack buffer zones as prey reservoirs. *Science*, *198*: 320–321.

Mech, L. D., and. Karns, P. D. (1977). *Role of the Wolf in a Deer Decline in the Superior National Forest*. U.S. Forest Service Research Paper NC-148, North Central Forest Experiment Station, St. Paul, MN.

Miller, B. J., Harlow, H. J., Harlow, T. S., Biggins, D., and Ripple, W. J. (2012). Trophic cascades linking wolves (Canis lupus), coyotes (Canis latrans), and small mammals. *Canadian Journal of Zoology*, 90: 70–78.

Newsome, T. M., and Ripple, T. M. (2015). A continental scale trophic cascade from wolves through coyotes to foxes. *Journal of Animal Ecology*, *84*: 49–59.

Ostfeld, R. S., and Keesing, F. (2000). Pulsed resources and community dynamics of consumers in terrestrial ecosystems. *Trends in Ecology and Evolution*, *15*: 232–237.

Pace, M. L., Cole, J. J., Carpenter, S. R., and Kitchell, J. F. (1999). Trophic cascades revealed in diverse ecosystems. *Trends in Ecology and Evolution*, *14*: 483–488.

Polis, G. A., and Strong, D. R. (1996). Food web complexity and community dynamics. *The American Naturalist*, *147*: 813–846.

Preisser, E. L., Bolnick, D. I., and Benard, M. F. (2005). Scared to death? The effects of intimidation and consumption in predator-prey interactions. *Ecology*, 86(2): 501–509.

Ripple, W. J., and Beschta, R. L. (2012). Trophic cascades in Yellowstone: The first 15 years after wolf reintroduction. *Biological Conservation*, *145*: 205–213.

Ripple, W. J., Estes, J. A., Beschta, R. L., Wilmers, C. C., Ritchie, E. G., Hebblewhite, M., Berger, J. Elmhagen, B., Letnic, M., Nelson, M. P., Schmitz, O. J., Smith, D. W., Wallach, A. D., and Wirsing, A. J. (2014). Status and ecological effects of the world's largest carnivores. *Science*, *343*: 1241484.

Ripple, W. J., Wirsing, A. J., Wilmers, C. C., and Letnic, M. (2013). Widespread mesopredator effects after wolf extirpation. *Biological Conservation*, *160*: 70–79.

Ritchie, E. G., Elmhagen, B., Glen, A. S., Letnic, M., Ludwig, G., and McDonald, R. A. (2012). Ecosystem restoration with teeth: What role for predators? *Trends in Ecology and Evolution*, *27*(5): 265–271.

Rooney, T. P., and Anderson, D. P. (2009). Are wolf-mediated trophic cascades boosting biodiversity in the Great Lakes region? In: Wydeven, A. P., Van Deelen, T. R., and Heske, E. J. (Eds.), *Recovery of Gray Wolves in the Great Lakes Region of the United States: An Endangered Species Success Story*. New York: Springer Science, 205–216.

Schmitz, O. J. (2005). Behaviors of predators and prey and links with population level processes. In: Barbosa, P., and Castellanos, I. (Eds.), *Ecology of Predator-Prey Interactions*. Oxford, UK: Oxford University Press, 256–278.

Schmitz, O. J., Krivan, V., and Ovadia, O. (2004). Trophic cascades: The primacy of trait-mediated indirect interactions. *Ecology Letters*, *7*: 153–163.

Schnurr, J. L., Canham, C. D., Ostfeld, R. S., and Inouye, R. S. (2004). Neighborhood analyses of small-mammal dynamics: Impacts on seed predation and seedling establishment. *Ecology*, *85*(3): 741–755.

Thiel, R. P. (1993). *The Timber Wolf in Wisconsin: The Death and Life of a Majestic Predator*. Madison, WI: University of Wisconsin Press.

Werner, E. E., and Peacor, S. D. (2003). A review of trait-mediated indirect interactions in ecological communities. *Ecology*, *84*: 1083–1100.

White, M. A. (2012). Long-term effects of deer browsing: Composition, structure and productivity in a northeastern Minnesota old-growth forest. *Forest Ecology and Management*, *269*: 222–228.

8 Linear corridors and predator movement dynamics in the Great Lakes Basin

Victoria M. Donovan and *Jesse N. Popp*

Introduction

The Great Lakes Basin is a highly developed region containing more than 30 million people (U.S. Environmental Protection Agency, 2016). In social-ecological systems where humans and nature coincide, it is important to understand how anthropogenic (non-natural) features interact with ecosystem processes. There is a growing understanding in the scientific community that linear anthropogenic landscape features such as roads and railways can substantially influence the structure and function of ecological systems (Fahrig and Rytwinski, 2009; Popp and Boyle, 2017; van der Ree et al., 2011; Van der Grift, 1999). Anthropogenic linear features are extensive across populated regions around the globe, and roads and railways make up the predominant linear features on the landscape in the Great Lakes Basin. Such features likely play a large role in shaping the terrestrial ecology of the region, yet we are only beginning to understand their ecological impacts and conservation implications for wildlife.

Roads are substantial threats to multiple species, ranging from plants (Lee et al., 2012) to mammals (Beyer et al., 2013; Northrup et al., 2012; Van der Grift, 1999), and are associated with wildlife mortality, changes in population persistence, and lower reproduction rates (Hebblewhite et al., 2003; Heinrichs et al., 2015; Hu et al., 2005; Whittington et al., 2011). Many species are known to avoid roads, particularly roads with higher traffic levels (Montgomery et al., 2012; Northrup et al., 2012). Such disturbances can create barriers to movement, causing habitat fragmentation and population isolation (Dyer et al., 2002; Laurance et al., 2004). In contrast, some roads with lower traffic volume act as movement corridors, allowing rapid colonization of new territories and altering trophic interactions (Courbin et al., 2014; Kunkel and Pletscher, 2001; Shannon et al., 2014; Whittington et al., 2011). Thus, the effects of roads on wildlife are complex and can alter ecosystem dynamics in multiple ways. Understanding how animals use roads and whether specific features alter wildlife use will help unravel the complexity associated with road–wildlife interactions.

Railway ecology is largely underrepresented in science, and there are few studies on the effects of railways on wildlife (Popp and Boyle, 2017). Railways can affect wildlife in ways similar to roads, including direct mortality, habitat loss, and habitat fragmentation (Dorsey et al., 2015; Van der Grift, 1999). Although train collision–induced fatalities are believed to be one of railways' primary impacts on wildlife (Becker and Grauvogel, 1991; Dorsey et al., 2015; Kušta et al., 2011), they also may alter wildlife movement behaviors (Dorsey et al., 2015). Little is known about the use of railways by wildlife; however, conservation and management goals could greatly benefit from a further understanding of railways' potential impacts and influences on ecosystem dynamics.

The impacts on wildlife of low-use linear features such as tertiary roads and railways are poorly understood. However, their potential to serve as movement corridors for wildlife suggest that they can have a large impact on ecosystem dynamics. We aim to understand the influence of tertiary roads and railways on two North American predators, the North American black bear (*Ursus americanus*) and wolves (*Canis lupus*, *Canis lycaon*, as well as hybrids of the two). Predators play a large role in shaping ecological systems, from altering prey community composition and distribution (e.g., Livingston et al., 2017; Seip, 1992) to far-reaching cascading effects on processes like disease dynamics, fire, carbon sequestration, invasive species, and biogeochemical exchanges in an ecosystem (Estes et al., 2011). Variance in social structures and behaviors in predators are important features influencing the dynamics of trophic interactions (Ritchie et al., 2012). Thus, understanding the relationship between low-use linear features and predators is highly relevant to understanding these features' potential impacts on ecosystem structure and processes. To examine how such features may influence predators' movement dynamics, we investigated seasonal railway use by wolves and black bears and identified habitat features that influence the intensity of wolf and black bear use of tertiary roads. Tertiary-road assessments in this chapter are paraphrased from Popp and Donovan (2016).

Statement of the issue

This study has two principal goals: The first is to document patterns in black bear and wolf occurrence along railways. The second is to determine fine-scale habitat features that influence bear and wolf occurrence on tertiary roads. This information will enhance our understanding of how low-use anthropogenic features can influence wildlife movement and behavior to support conservation and mitigation strategies in the Great Lakes Basin.

Methods

We conducted our study in the northeastern region of the Great Lakes Basin, approximately 30 kilometers (19 miles) south of the city of Sudbury, Ontario. Our study site near the abandoned Burwash Industrial Farm hosts thousands of hectares of open abandoned fields interspersed among conifer-hardwood forest (Chambers et al., 1996) and multiple water bodies (lakes, rivers, and wetlands). There are two primary roads (high-use highways) near the Burwash region, along with multiple tertiary roads (unpaved, low-use roads) used primarily for recreation and logging activities. The Canadian National Railway runs directly through our study area. Two large predators dominate this region: black bears and wolves. Black bear density estimates are one black bear every 4 square kilometers (1.5 square miles) (Ontario Ministry of Natural Resources [OMNR], 2014), while wolf density estimates are one wolf every 30 to 50 square kilometers (12 to 20 square miles) (Kittle et al., 2008). White-tailed deer (*Odocoileus virginianus*), elk (*Cervus elaphus*), moose (*Alces alces*), beaver (*Castor canadensis*), and muskrat (*Ondatra zibethicus*) are all common prey for these predators in our study area.

At the beginning of July 2015, 10 trail cameras were placed along a 20-kilometer (12-mile) stretch of railway bisecting the Burwash region. Cameras were placed every 2 kilometers (1.2 miles), attached to trees facing perpendicular to the railway. Cameras were set to record bursts of three images when motion was detected. However, image recordings were set to stop after five minutes of the initial burst to prevent excessive photography of trains. Camera batteries and memory cards were replaced every one to three months, and cameras were taken down at the end of June 2016. All photos showing a discernable image of a black bear or wolf were recorded. Photos taken in the same burst were merged into a single recorded sighting. We ensured photos were at least one hour apart for individuals on the same camera to avoid repetitive recordings of the same animal walking back and forth in a given area. To determine if we were capturing more than one individual over the course of our assessment, we examined photos for distinguishing markers such as face and body markings, size, and stature. We also determined whether wolves or black bear were walking along the railway (parallel), suggesting use as a movement corridor, or crossing the railway. Photos were divided into seasons (spring, summer, fall, or winter) to identify whether seasonal trends in railway occurrence existed for both species.

To document wolf and bear occurrence along tertiary roads, we created 32 road transects and recorded the location of bear and wolf scat within each transect between May to the end of August for a three-year period (2013–2015). Road transects were created by dividing tertiary roads into 500-meter (547-yard) segments, separated by at least 1 kilometer (0.6 miles) if on the same continuous stretch of road. All transects were deemed to be spatially independent from one another. To detect scat on tertiary roads, roads were visually scanned by the same observer every two weeks. Once recorded, scat was cleared from roads to prevent recounting during the next assessment.

To identify fine-scale habitat features that may influence wolf and bear use of roads, we collected these fine-scale habitat features within each road transect during the last year of sampling (2015): road width, road cover type (percent herbaceous, gravel, dirt), the percent of vegetation cover types along the roadside (herbaceous, shrub, wetland, forest), and canopy cover. Both road width and canopy cover were thought to potentially influence the level of vulnerability an animal feels while using a roadway, while vegetation cover was associated with forage availability. Black bears have been found to select areas with higher abundances of food items, such as grasses, sedges, and fruit typically found in wetlands and open canopy areas (Costello, 1992; Costello and Sage, 1994; Fecske et al.,

2002; Raine and Kansas, 1990; Romain et al., 2013). Wolves, on the other hand, generally select areas with a higher number of prey items, and thus their selection of foraging locations is driven by prey habitat selection patterns (Lesmerises et al., 2012). Moose, white-tailed deer, and elk select wetlands, grasslands, and shrub-rich areas, while smaller prey items, like muskrat and beaver, prefer wetland areas (Anderson et al., 2005; Massé and Côté, 2012; Street et al., 2015). We also measured the straight-line distance from the center of each road segment to the nearest primary road because both wolves and bears have been shown to avoid high-traffic areas (Gurarie et al., 2011; Van Manen et al., 2012).

We used Akaike's Information Criterion (AICc), a model selection technique for small sample sizes, to understand what variables most influenced predator road use. To do this, we created six models with different combinations of road features associated with foraging opportunities and anthropogenic disturbance for wolves and black bears. We then used AICc to rank each model based on which habitat features best explained predator tertiary road use.

Findings

In total, we recorded 222 wolf occurrences and 43 black bear occurrences along the 20 kilometers (12.5 miles) of railway. Based on characteristic markers, we identified at least three individual wolves and at least four individual bears using the railway. However, it was much more difficult to identify unique individual characteristics on black bears than on wolves. Wolves tended to use railways much more often than black bears, even though black bears occur at higher densities in our study region. All images of wolves along the railway were of individuals walking parallel with the rail line (Figure 8.1). For black bears, 88 percent (n = 38) of occurrences recorded individuals walking along the railway (parallel)

Figure 8.1 Wolf walking along a railway line.

Source: Victoria M. Donovan and Jesse N. Popp

Figure 8.2 Bear walking next to railroad tracks.

Source: Victoria M. Donovan and Jesse N. Popp

Figure 8.3 Bear crossing railroad tracks.

Source: Victoria M. Donovan and Jesse N. Popp

(Figure 8.2), while 12 percent (n = 5) of black bear occurrences consisted of individuals crossing or entering/exiting the railway (Figure 8.3). Wolves generally used the railway tracks most frequently in the summer and fall, while bears frequented the tracks most commonly in the summer (Table 8.1). Black bears

Table 8.1 Total number of seasonal bear and wolf occurrences on a 20-kilometer (12.5-mile) stretch of railway during one year (2015–2016)

	Spring	*Summer*	*Fall*	*Winter*
Black Bear	13	25	5	0
Wolf	27	72	73	50

were five times more likely to travel on the railway tracks in the summer than in the fall, while wolves were almost three times more likely to be photographed on the railway tracks in the summer and fall compared to spring. We recorded no black bear occurrences on railway tracks during the winter.

For tertiary roads, we recorded 73 wolf and 97 bear scats at our sample transects during our study period. Two features best explained both bear and wolf use of tertiary roads: the distance of the road to a highway and the width of the road they were using. Bears tended to use narrower roads that are farther away from highways. Wolf tertiary road use was influenced by an interaction between road width and distance to highways. Wolves generally used narrower tertiary roads near the highway. However, as they went farther from the highway, the effect of road width on their road-use decisions became smaller. Models containing vegetation features were generally the lowest-ranking models for both species, suggesting they explained bear and wolf occurrence on tertiary roads the least of all proposed models.

Discussion and conclusions

These findings highlight the importance of low-use linear features to major predators in the Great Lakes Basin. In particular, both black bear and wolf occurrence on tertiary roads was positively associated with the distance to highways, where both species were more likely to occur on tertiary roads further from high-traffic areas. Both bears and wolves have previously been shown to avoid high-traffic areas (Gurarie et al., 2011; Van Manen et al., 2012). Similarly, both predators are believed to experience increased levels of stress when closer to anthropogenic features (Creel et al., 2002; Ditmer et al., 2015). Their preference for narrower roads may relate to a sense of security. The fact that road width became less important as wolves moved farther from highways could suggest that they feel safer on narrow roads when nearer to high-traffic areas. Multiple other factors must be considered in relation to this hypothesis, however. For example, vehicles such as ATVs and logging machinery may travel more on wider tertiary roads. In addition, access to tertiary roads closer to highways may be easier for recreational road users. Thus, further investigation into the relationship between tertiary road activity levels and bear and wolf road use is necessary. That said, these findings support studies suggesting that black bears and wolves may tend to avoid areas near high-traffic roads (Gurarie et al., 2011; Van Manen et al., 2012).

Both bears and wolves move predominantly parallel along railway tracks, suggesting their use of railways as movement corridors. As with low-traffic roads that may facilitate predators' movement (Gurarie et al., 2011; Zimmermann et al., 2014), railways could support easy travel among different regions of an animal's range (Van der Grift, 1999). Large mammals are known to use railways, especially in winter when snow conditions impede movement off-rail (Van der Grift, 1999). However, wolves used the railway substantially in all seasons but less so in winter when one would expect increased use solely related to ease of movement. On the other hand, the failure to find black bears using railways in the winter is likely associated with their winter hibernation patterns (Kolenosky and Strathearn, 1987). In addition to providing easy travel routes, predators may identify railway corridors as ideal areas to scavenge train-kills (Wells et al., 1999). Trains cause collision-induced wildlife fatalities among many species and thus are likely to provide scavenging opportunities (Andreassen et al., 2005; Dasgupta and Ghosh, 2015; Gundersen and Andreassen, 1998; Iosif, 2012; Kusta et al., 2014; Van der Grift, 1999). For example, wolf and coyote train-collision deaths have been linked to locations with train-killed ungulate carcasses (Van der Grift, 1999). In addition to scavenging opportunities, railways produce other forage opportunities, especially for bears, when grain spills occur (Wells et al., 1999). Such spills could similarly attract other prey species such as ungulates, creating additional hunting opportunities for predators. Regardless of the reasons behind black bear and wolf movement along railways, railways appear to be important travel corridors and may therefore encourage rapid population expansion, lead to train-collision fatalities, or alter trophic interactions among predator and prey species.

Forage availability has also been linked to predator road use (Roever et al., 2008). However, this study did not find strong evidence that black bears or wolves use tertiary roads for the vegetation features that offer forage opportunities. For both species, road width and distance to primary roads were the most important features influencing species occurrence on tertiary roads. Moreover, although black bears and wolves demonstrate different foraging strategies, such as opportunistic predator versus active predator (Bastille-Rousseau et al., 2011; Lesmerises et al., 2012), we found strong similarities in factors driving their occurrence on these roads. This suggests that forage is not likely the primary motivator. Instead, we suggest they use tertiary roads, especially those that are more secluded with less traffic, as movement corridors, allowing them to move quickly across their range. Research has shown that wolves select locations that facilitate travel (Kunkel and Pletscher, 2001), and multiple studies have suggested that tertiary roads facilitate wolves' movement (Gurarie et al., 2011; Zimmermann et al., 2014). Dickie et al. (2016) found that wolves moved faster and farther while using linear features. With our scat sampling method, we were unable to determine whether black bears or wolves were moving parallel to roads or using them as crossing points. Thus, GPS collaring would be useful to test this hypothesis.

As the Great Lakes Basin continues to develop, it becomes ever more important to understand the complex relationships between wildlife species and anthropogenic structures. Our results have twofold implications for the conservation of terrestrial wildlife in this region. First, our findings highlight two features that remain underrepresented in our basic and applied research of non-natural features that can dramatically alter species movements and interactions. Second, our results suggest potential features that managers can manipulate (road width, distance to highways) to facilitate predators' movement in fragmented landscapes or alternatively limit their movement into regions where rapid colonization of predators is undesirable or predation pressures are too high. That said, the density of roads and railways in our study region is far less than in some other regions of the Great Lakes Basin. Thus, promoting predator movement along railways and tertiary roads will likely be more limited in regions with high densities of high-traffic roads. How these features shape trophic interactions, alter wildlife movement, and influence colonizations and extinctions are among the many questions to explore as we deepen our understanding of Great Lakes Basin ecology.

References

Anderson, D. P., Turner, M. G., Forester, J. D., Zhu, J., Boyce, M. S., Beyer, H., and Stowell, L. (2005). Scale-dependent summer resource selection by reintroduced elk in Wisconsin, USA. *Journal of Wildlife Management*, 69: 298–310.

Andreassen, H. P., Gundersen, H., and Storaas, T. (2005). The effect of scent-marking, forest clearing, and supplemental feeding on moose – train collisions. *Journal of Wildlife Management*, 69: 1125–1132.

Bastille-Rousseau, G., Fortin, D., Dussault, C., Courtois, R., and Ouellet, J. (2011). Foraging strategies by omnivores: Are black bears actively searching for ungulate neonates or are they simply opportunistic predators? *Ecography*, *34*: 588–596.

Becker, E. F., and Grauvogel, C. A. (1991). Relationship of reduced train speed on moose-train collisions in Alaska. *Alces*, *27*: 161–168.

Beyer, H. L., Ung, R., Murray, D. L., and Fortin, M. (2013). Functional responses, seasonal variation and thresholds in behavioural responses of moose to road density. *Journal of Applied Ecology*, *50*: 286–294.

Chambers, B., Legasy, K., Bentley, C. V., LaBelle-Beadman, S., and Thurley, E. (1996). *Forest Plants of Central Ontario*. Canada: Lone Pine Publishing.

Costello, C. M. (1992). *Black bear habitat ecology in the central Adirondacks as related to food abundance and forest management*. M.S. Thesis. State University of New York, College of Environmental Science and Forestry, Syracuse, p. 165.

Costello, C. M., and Sage Jr., R. W. (1994). Predicting black bear habitat selection from food abundance under 3 forest management systems. *Bears: Their Biology and Management*, 9: 375–387.

Courbin, N., Fortin, D., Dussault, C., and Courtois, R. (2014). Logging-induced changes in habitat network connectivity shape behavioral interactions in the wolf – caribou – moose system. *Ecological Monographs*, *84*: 265–285.

Creel, S., Fox, J. E., Hardy, A., Sands, J., Garrott, B., and Peterson, R. O. (2002). Snowmobile activity and glucocorticoid stress responses in wolves and elk. *Conservation Biology*, *16*: 809–814.

Dasgupta, S., and Ghosh, A. K. (2015). Elephant – railway conflict in a biodiversity hotspot: Determinants and perceptions of the conflict in northern West Bengal, India. *Human Dimensions of Wildlife*, *20*: 81–94.

Dickie, M., Serrouya, R., McNay, R. S., and Boutin, S. (2016). Faster and farther: Wolf movement on linear features and implications for hunting behaviour. *Journal of Applied Ecology* 54: 253–263.

Ditmer, M., Garshelis, D., Noyce, K., Laske, T., Iaizzo, P., Burk, T., Forester, J., and Fieberg, J. (2015). Behavioral and physiological responses of American black bears to landscape features within an agricultural region. *Ecosphere*, *6*: 1–21.

Dorsey, B., Olsson, M., Rew, L. J., van der Ree, R., Smith, D., and Grilo, C. (2015). Ecological effects of railways on wildlife. In: van der Ree, R., Smith, D. J., and Grilo, C. (Eds.), *Handbook of Road Ecology*. Oxford, UK: John Wiley & Sons, 219–227.

Dyer, S. J., O'Neill, J. P., Wasel, S. M., and Boutin, S. (2002). Quantifying barrier effects of roads and seismic lines on movements of female woodland caribou in northeastern Alberta. *Canadian Journal of Zoology*, *80*: 839–845.

Estes, J. A., Terborgh, J., Brashares, J. S., Power, M. E., Berger, J., Bond, W. J., Carpenter, S. R., Essington, T. E., Holt, R. D., and Jackson, J. B. (2011). Trophic downgrading of planet Earth. *Science*, *333*: 301–306.

Fahrig, L., and Rytwinski, T. (2009). Effects of roads on animal abundance: An empirical review and synthesis. *Ecology and Society*, *14*: 21.

Fecske, D. M., Barry, R. E., Precht, F. L., Quigley, H. B., Bittner, S. L., and Webster, T. (2002). Habitat use by female black bears in western Maryland. *Southeastern Naturalist*, *1*: 77–92.

Gundersen, H., and Andreassen, H. P. (1998). The risk of moose Alces collision: A predictive logistic model for moose-train accidents. *Wildlife Biology*, *4*: 103–110.

Gurarie, E., Suutarinen, J., Kojola, I., and Ovaskainen, O. (2011). Summer movements, predation and habitat use of wolves in human modified boreal forests. *Oecologia*, *165*: 891–903.

Hebblewhite, M., Percy, M., and Serrouya, R. (2003). Black bear (Ursus americanus) survival and demography in the Bow Valley of Banff National Park, Alberta. *Biological Conservation*, *112*: 415–425.

Heinrichs, J. A., Bender, D. J., Gummer, D. L., and Schumaker, N. H. (2015). Effects of landscape and patch-level attributes on regional population persistence. *Journal for Nature Conservation*, *26*: 56–64.

Hu, Z., Yu, C., Xu, H., and Wang, Y. (2005). Ecological effects of roads on terrestrial animals. *Chinese Journal of Ecology*, *24*: 433–437.

Iosif, R. (2012). Railroad-associated mortality hot spots for a population of Romanian Hermann's tortoise (Testudo Hermanni Boettgeri): A gravity model for railroad-segment analysis. *Procedia Environmental Sciences*, *14*: 123–131.

Kittle, A. M., Fryxell, J. M., Desy, G. E., and Hamr, J. (2008). The scale-dependent impact of wolf predation risk on resource selection by three sympatric ungulates. *Oecologia*, *157*: 163–175.

Kolenosky, G. B., and Strathearn, S. M. (1987). Winter denning of black bears in east-central Ontario. *Bears: Their Biology and Management*, *7*: 305–316.

Kunkel, K., and Pletscher, D. H. (2001). Winter hunting patterns of wolves in and near Glacier National Park, Montana. *Journal of Wildlife Management*, *65*: 520–530.

Kusta, T., Hola, M., Keken, Z., Jezek, M., Zika, T., and Hart, V. (2014). Deer on the railway line: Spatiotemporal trends in mortality patterns of roe deer. *Turkish Journal of Zoology*, *38*: 479–485.

Kušta, T., Ježek, M., and Keken, Z. (2011). Mortality of large mammals on railway tracks. *Scientia Agriculturae Bohemica*, *42*: 12–18.

Laurance, S. G. W., Stouffer, P. C., and Laurance, W. F. (2004). Effects of road clearings on movement patterns of understory rainforest birds in Central Amazonia. *Conservation Biology*, *18*: 1099–1109.

Lee, M. A., Davies, L., and Power, S. A. (2012). Effects of roads on adjacent plant community composition and ecosystem function: An example from three calcareous ecosystems. *Environmental Pollution*, *163*: 273–280.

Lesmerises, F., Dussault, C., and St-Laurent, M. -H. (2012). Wolf habitat selection is shaped by human activities in a highly managed boreal forest. *Forest Ecology and Management, 276*: 125–131.

Livingston, G., Fukumori, K., Provete, D. B., Kawachi, M., Takamura, N., and Leibold, M. A. (2017). Predators regulate prey species sorting and spatial distribution in microbial landscapes. *Journal of Animal Ecology*, 86: 501–510.

Massé, A., and Côté, S. D. (2012). Linking habitat heterogeneity to space use by large herbivores at multiple scales: From habitat mosaics to forest canopy openings. *Forest Ecology and Management, 285*: 67–76.

Montgomery, R. A., Roloff, G. J., and Millspaugh, J. J. (2012). Importance of visibility when evaluating animal response to roads. *Wildlife Biology, 18*: 393–405.

Northrup, J. M., Pitt, J., Muhly, T. B., Stenhouse, G. B., Musiani, M., and Boyce, M. S. (2012). Vehicle traffic shapes grizzly bear behaviour on a multiple-use landscape. *Journal of Applied Ecology, 49*: 1159–1167.

OMNR. (2014). *Black Bear Ecological Zones (BBEZ), Black Bear Density Map*. Available at www.ontario.ca/document/bear-density-map

Popp, J., and Boyle, S. (2017). Railway ecology: Underrepresented in science? *Basic and Applied Ecology, 19*: 84–93. doi: 10.1016/j.baae.2016.11.006

Popp, J. N., and Donovan, V. M. (2016). Fine-scale tertiary-road features influence wildlife use: A case study of two major North American predators. *Animal Biology*, 66: 229–238.

Raine, R. M., and Kansas, J. L. (1990). Black bear seasonal food habits and distribution by elevation in Banff National Park, Alberta. *Bears: Their Biology and Management*, 8: 297–304.

Ritchie, E. G., Elmhagen, B., Glen, A. S., Letnic, M., Ludwig, G., and McDonald, R. A. (2012). Ecosystem restoration with teeth: What role for predators? *Trends in Ecology and Evolution, 27*: 265–271.

Roever, C., Boyce, M., and Stenhouse, G. (2008). Grizzly bears and forestry: II: Road vegetation and placement as an attractant to grizzly bears. *Forest Ecology and Management, 256*: 1253–1261.

Romain, D. A., Obbard, M. E., and Atkinson, J. L. (2013). Temporal variation in food habits of the American black bear (Ursus americanus) in the boreal forest of northern Ontario. *Canadian Field Naturalist, 127*: 118–130.

Seip, D. R. (1992). Factors limiting woodland caribou populations and their interrelationships with wolves and moose in southeastern British Columbia. *Canadian Journal of Zoology, 70*: 1494–1503.

Shannon, G., Cordes, L. S., Hardy, A. R., Angeloni, L. M., and Crooks, K. R. (2014). Behavioral responses associated with a human-mediated predator shelter. *PLOS ONE*, 9: e94630.

Street, G. M., Rodgers, A. R., and Fryxell, J. M. (2015). Mid-day temperature variation influences seasonal habitat selection by moose. *Journal of Wildlife Management, 79*: 505–512.

U.S. Environmental Protection Agency. (2016). *Great Lakes Facts and Figures*. Available at www.epa.gov/greatlakes/great-lakes-facts-and-figures

Van der Grift, E. (1999). Mammals and railroads: Impacts and management implications. *Lutra, 42*: 77–98.

van der Ree, R., Jaeger, J. A. G., van der Grift, E., and Clevenger, A. (2011). Effects of roads and traffic on wildlife populations and landscape function: Road ecology is moving toward larger scales. *Ecology and Society*, 16: 48.

Van Manen, F. T., Mccollister, M. F., Nicholson, J. M., Thompson, L. M., Kindall, J. L., and Jones, M. D. (2012). Short-term impacts of a 4-lane highway on American black bears in eastern North Carolina. *Wildlife Monographs, 181*: 1–35.

Wells, P., Woods, J. G., Bridgewater, G., and Morrison, H. (1999). Wildlife mortalities on railways: Monitoring methods and mitigation strategies. *Proceedings of the Third International Conference on Wildlife Ecology and Transportation*. September 13–16, Missoula, MT. Florida Department of Transportation report FL-ER-73-99.

Whittington, J., Hebblewhite, M., DeCesare, N. J., Neufeld, L., Bradley, M., Wilmshurst, J., and Musiani, M. (2011). Caribou encounters with wolves increase near roads and trails: A time-to-event approach. *Journal of Applied Ecology*, *48*: 1535–1542.

Zimmermann, B., Nelson, L., Wabakken, P., Sand, H., and Liberg, O. (2014). Behavioral responses of wolves to roads: Scale-dependent ambivalence. *Behavioural Ecology*, *25*: 1353–1364.

Part III

Pollution, climate change, and invasive species

9 Toxicants in the Great Lakes

Living with a toxic legacy while managing for chemicals of emerging concern

Dalma Martinović-Weigelt, Jane R. Feely, and *Heiko L. Schoenfuss*

Introduction

The earliest record of anthropogenic contamination of the Great Lakes comes from records of the Keweenaw Peninsula in Lake Superior, with evidence that prehistoric lead pollution from copper production occurred around 3,000 years ago (Pompeani et al., 2013). Rapid industrialization and new technologies of the twentieth century left a lasting mark on the Great Lakes. Until the 1970s, persistent organic pollutants, heavy metals, and contaminants from industrial and agricultural uses entered the lakes via intentional discharges, as well as from surface and atmospheric deposition. Much initial attention focused on phosphorus loading and thermal pollution, but the toxic chemical problem gained attention in the late 1960s.

When Ohio's polluted Cuyahoga River, which drains into Lake Erie, caught fire in June 1969 – after at least a dozen prior burning incidents – it brought attention to the widespread contamination of North American waters and catalyzed enactment of pollution control measures. Creation of the Great Lakes Water Quality Agreement (GLWQA) in 1972 committed the United States and Canada to restore and protect the lakes. It built on efforts of the International Joint Commission to prevent and resolve disputes between the two countries.

Notably, widespread public concern over the gravity of Great Lakes pollution led to a major overhaul of the Federal Water Pollution Control Act of 1948 and gave birth to the Clean Water Act (CWA) in 1972. The CWA established a basic framework for the regulation of pollutants in aquatic systems and remains the primary federal law governing water quality in the United States.

From "Burn On" by Randy Newman (1972)

> There's an oil barge winding
> Down the Cuyahoga River
> Rolling into Cleveland to the lake
> Cleveland, city of light, city of magic
> Cleveland, city of light, you're calling me
> Cleveland, even now I can remember
> 'Cause the Cuyahoga River

goes smokin' through my dreams
Burn on, big river, burn on
Burn on, big river, burn on
Now the Lord can make you tumble
Lord can make you turn
The Lord can make you overflow
But the Lord can't make you burn

Over the last 60 years, the Great Lakes have served as a stage for the discovery of contaminants responsible for deteriorating ecological health and catalyzed development of innovative monitoring, risk assessment, and regulatory approaches for contaminant and ecosystem management.

Statement of the problem

The adverse effects of most manufactured chemicals were not well characterized in the mid-twentieth century. However, monitoring programs resulting from the GLWQA generated evidence pointing to chemicals as the potential cause of widely observed health problems. Typically, severe adverse effects such as fish and wildlife population declines, gross malformations, and developmental disorders were first observed in wild populations. Through forensic and follow-up analyses, researchers identified chemical drivers of toxic responses. This chapter highlights well-known and widely studied environmental issues associated with persistent, long-banned ("legacy") contaminants lingering in the Great Lakes and describes how environmental policies instituted in the 1970s have contributed to water quality improvements. Lastly, the chapter evaluates the most recent challenges there, namely contaminants of emerging concern (CECs).

Methods

This chapter draws on well-known studies and reviews studies of occurrence and effects of CECs. It also describes ongoing work by the CEC Great Lakes Restoration Initiative project, in which the authors participate.

Findings

Forensic approaches to identify chemical drivers of toxicity: success stories

One story illustrating the complexities of this forensic/retrospective approach to identify chemical drivers of adverse outcomes is the discovery of impaired hatchability and widespread deformities in colonial fish-eating birds, including subcutaneous and cardiac edema and beak and skeletal malformations, in the early 1970s. This phenomenon is called Great Lakes embryo mortality, edema, and deformities syndrome (GLEMEDS). It was well-known then that such a suite of adverse effects existed in domestic poultry inadvertently exposed to dioxins, but due to the poorly characterized environmental chemistry of dioxin and dioxin-related compounds and the lack of sufficiently sensitive analytical techniques, not until the early 1980s could scientists confirm the presence of dioxin-like compounds in Great Lakes' bird eggs (Gilbertson et al., 1983). Only in the early 1990s – well after implementation of ambient water criteria for dioxins – could researchers confidently ascribe GLEMEDS to dioxin-like compounds in the lakes (Gilbertson et al., 1991). Dioxin-like organic compounds such as polychlorinated dibenzodioxins, including 2,3,7,8-tetrachlorodibenzo-p-dioxin (TCDD), are by-products of organochloride manufacturing.

Another, albeit rare, success story linking adverse effects in wild populations with high certainty to a particular contaminant and/or group of contaminants involves lake trout (*Salvelinus namaycush*). Their decline in Lake Ontario started in the 1930s, and they were almost extinct by the 1960s. At the time, it was widely accepted that overfishing, habitat destruction, and lamprey predation caused the population collapse. Despite intense efforts to rehabilitate the population by stocking, recovery was failing. Lake trout embryos – offspring of wild-caught adults artificially spawned in restocking programs – showed a high incidence of "blue-sac" disease. That disease is characterized by cytochrome P450 (CYP1A) induction, yolk sac edema, hemorrhaging, craniofacial deformities, circulatory failure, and impaired development (Spitsbergen et al., 1991). Studies demonstrated that comparable pathologies can be induced by chemical agonists of aryl hydrocarbon receptors such as dioxin-like compounds and some polychlorinated biphenyls (PCBs). Finally, researchers used exposure modeling and direct measurement of chemicals in dated sediment cores and wild lake trout eggs to establish that toxicity thresholds were likely exceeded (Wright and Tillit, 1999). In the end, chemical occurrence and epidemiological data, combined with laboratory experimentation and modeling, provided strong evidence that TCDD-associated mortality of lake trout sac fry could adversely affect the population without the contribution of other chemical or nonchemical stressors (Cook et al., 2003).

While skepticism initially met these forensic epidemiological approaches (Gilbertson, 1997), they eventually led to the use of rule-based, weight-of-evidence approaches that evaluate the strength of association between contaminants and adverse outcomes. Typically such analyses evaluate biological probability; time order, meaning determination of whether chemical releases precede adverse effects; strength and consistency of association across locations, species, and research teams; and predictive performance (Tillitt et al., 2008).

Other advances

Other major contributions to environmental toxicology developed from Great Lakes' monitoring programs and research initiatives. They include advanced understanding of the sources, occurrence, and fate of contaminants, not just identification of adverse effects and the chemicals responsible. For example, studies determined that many chemicals persist in the environment and that lake sediments can serve as their "sinks." Another major discovery was the importance of atmospheric deposition of contaminants in the Great Lakes, which has been demonstrated for polycyclic hydrocarbons (McVeety and Hites, 1988) and trace elements, including mercury (Winchester et al., 1971). Finally, researchers learned that the major route of human exposure to persistent contaminants is through contaminated food, particularly fish (Cordle et al., 1978). Fish surveys showed that most species, including commercially important salmon and trout, contained PCBs and mercury; tissue concentrations of these contaminants correlated positively with fish size and age. Concentrations of PCBs and mercury were highest in predator species, indicating their potential to bioconcentrate through the food web. Finally, studies established that many of those who ate Great Lakes fish exceeded the recommended consumption of mercury (Turney et al., 1971) and PCBs (Cordle et al., 1978). Thus government agencies banned commercial and sport fishing in many areas, leading to economic losses and lawsuits.

Reducing persistent toxic chemicals: "The Dirty Dozen"

As evidence mounted that contaminants could persist, in 1978 the United States and Canada committed themselves to monitoring and reducing persistent toxic chemicals ("second GLWQA"). As a result, contaminants in the lakes were reduced during the late 1970s and 1980s, largely due to regulation of production and application (Hornbuckle et al., 2006). In 1997, both countries adopted the Great Lakes Binational Toxics Strategy intended to "virtually eliminate" 12 chemicals ("The Dirty Dozen") in the Basin over a 10-year period. Many targeted chemicals were persistent organic pollutants, or POPs. The strategy acknowledged that governments alone cannot achieve that goal, and advocated broad involvement of all stakeholders and use of incentive-based actions. As a result, much progress has occurred: discharges of many POPs almost entirely ceased, and significant reductions of POPs occurred in sediments, fish tissues, and bird eggs. Some reductions exceed 90 percent (Marvin et al., 2004; Carlson et al., 2010; Solla et al., 2016).

Importantly, however, reductions and rates of reduction vary from lake to lake, largely related to land use, historical contamination, and differences in lake morphology and ecology such as depth and water residence time (Marvin et al., 2004). Finally, many Areas of Concern (AOCs) remain under GLWQA. There were 27 AOCs in the Basin as of December 2016, and in several cases the U.S. Environmental Protection Agency has identified legacy contaminants as the problem.

The future of POPs

While discharges of many POPs declined, analyses of fish tissues indicate the rate of decline has slowed significantly. For example, Carlson et al. (2010) reported that half-lives for oxychlordane, PCBs, DDT, and its metabolites are around 15 to

30 years; in the 1980s, they were only 3 to 6 years. The researchers predicted that legacy contaminants would continue to pose health concerns for decades unless remaining sources are identified and eliminated.

Since the 1990s, new groups of POPs have emerged as contaminants of interest, including fire retardants like polybrominated biphenyls, or PBDEs (Carlson et al., 2010). Government-mandated and voluntary removal from the market (2004–2013) reduced some PBDEs (Venier et al., 2015), but studies show the emergence of new flame retardants that were developed to replace PBDEs. For example, replacements such as tetrabrominated benzoates and phthalates and organophosphate esters are now being detected in air samples on the shores of the Great Lakes (Venier et al., 2015).

Contaminants of emerging concern

Contaminant research and monitoring efforts were reinvigorated in 2010 by a major investment from the White House through the Great Lakes Restoration Initiative (GLRI). The goal was to protect and restore the integrity of the Basin's ecosystem. The GLRI Action Plan (2010) identified toxic substances as one of five major focus areas. It called for clean-up and restoration of AOCs and historically contaminated sites to reduce sources of legacy contaminants. It also identified CECs as ecosystem threats and called for identification of significant sources of new contaminants. This was to be achieved "through robust surveillance as well as laboratory and field studies, in order to devise and implement effective control strategies" (Great Lakes Restoration Initiative Action Plan, 2010: 21). Similarly, the 2009 Great Lakes Chemicals of Emerging Concern Advisory Group report to the International Joint Commission identified a need to determine whether CECs jeopardize the ecosystem and human health.

It's complicated: what's old can be new, what's not persistent can persist

CECs encompass a broad range of chemicals, many of which have not been monitored or tested. There is no universally accepted definition. Current definitions encompass some chemicals that are well characterized and regulated and some that are not. They can be defined based on chemical structure, such as steroids, or on interaction with a biological entity, such as endocrine disruptors (Daughton, 2005). Furthermore, definitions can include not only new products but also those regulated for decades (Daughton, 2005; Sauvé and Desrosiers, 2014). For example, discoveries of new, subtle, unanticipated effects such as the disruption of hormonal systems by well-known legacy contaminants such as PCBs and mercury have led to categorizing them as CECs (Denham et al., 2005).

CECs include pharmaceuticals, fragrances, flame retardants, and pesticides. Most definitions include chemicals that are poorly characterized in occurrence, fate, and effects or of concern due to their "pseudo-persistence." Many CECs degrade relatively quickly – in hours or days, in contrast to years for POPs – but continual input into the environment means they are almost always present (Daughton, 2005). Unlike past contaminants whose lethality and overt adverse effects were major concerns, sublethal, subtle effects are the main sources of concern now. Endocrine-disrupting chemicals are among the best-studied groups of CECs. Their hallmarks are biological activity at very low concentrations, "pseudo-persistence," and lack of overt toxicity at environmentally relevant concentrations (Tyler et al., 1998).

Occurrence and sources of CECs

The veil of uncertainty surrounding the occurrence and sources of CECs slowly lifted at the beginning of the twenty-first century as advancements in analytical chemistry techniques allowed for increased reconnaissance. A 2009 review of scientific literature by the Great Lakes Chemicals of Emerging Concern Advisory Group reported detection of the following CECs: synthetic musks; fluorinated surfactants; brominated diphenyl ethers and other flame retardants; alkylphenol ethoxylates; chlorinated paraffins; phthalates; pesticides and pharmaceuticals; veterinary drugs; and personal care products. By 2016, additional studies focusing on the Great Lakes watershed firmly established the widespread occurrence of pharmaceuticals, personal care products, and other CECs in the lakes and their tributaries (Blair et al., 2013; Baldwin et al., 2016).

Sources of CECs and the fate of these compounds are also becoming clearer. Given the region's agricultural history, combined with intense industrialization of nearby urban centers, it is unsurprising that the occurrence of many CECs is linked to agricultural and urban runoff, as well as to controlled discharge from wastewater treatment plants (WWTPs) (Corsi et al., 2001; Barber et al., 2015; Baldwin et al., 2016). Steroid hormones and pharmaceuticals secreted through human urine and feces mix with personal care products, such as shampoos and detergents, and pass through WWTPs on their way to waterways (Barber et al., 2015). Agricultural and urban runoffs often contain mixtures of manure from fields and pet waste from urban parks, in addition to herbicides and pesticides. During snowmelt or rain, these CECs join legacy contaminants washing into Great Lakes tributaries (e.g., estrogenic chemicals; Martinović-Weigelt et al., 2013). As runoff is notoriously difficult to study because of its spatial and temporal unpredictability, much CEC source research focuses on the human waste stream passing through WWTPs. Efforts to improve water quality have paradoxically resulted in WWTP discharge becoming the most concentrated and consistent source of CECs to many of the lakes and their tributaries. WWTPs were intentionally designed to remove nutrients from municipal wastewater but not CECs. As such, WWTPs do not fully remove CECs and sometimes can transform CECs into more toxic metabolites (Stalter et al., 2010).

Effects of CECs: a great unknown

Among the triad of occurrence, fate, and effects, the least is known about the effects of CEC exposure on aquatic life. Studies show that many CECs have sublethal effects – meaning they are not overtly toxic but can negatively affect organisms in environmental concentrations – by changing behavior, physiology, or even the anatomy of exposed organisms. Exposure, especially to mood-altering pharmaceuticals and endocrine-active chemicals, may change feeding behavior (Brodin et al., 2013), increase response time to predators (Painter et al., 2009), or reduce males' ability to acquire and defend nesting territory (Martinovic et al., 2007). Following exposure to potent estrogens or androgens commonly found in domestic and industrial effluents, male fish may become demasculinized, and some fish may become hermaphroditic (Hirai et al., 2006; Vajda et al., 2011). These behavioral, physiological, and anatomical changes can harm fish populations by reducing reproduction (Kidd et al., 2007).

Despite these well-established effects, it is difficult to ascertain how much CECs affect Great Lakes fishes that may also be assaulted by legacy contamination, eutrophication, habitat loss, and competition from non-native species. Not only is it difficult to differentiate the effects of legacy contaminants from those of CECs, but even nonchemical stressors can initiate biological responses indistinguishable from those initiated by CECs. For example, effects of the hypoxia (low oxygen conditions) on ovulation and reproductive genes and hormones in fish closely resemble effects of endocrine-disrupting chemicals (Hala et al., 2012). Nevertheless, efforts are underway to study CEC impacts on fishes by combining the expertise of governmental, industrial, and academic sectors as part of the GLRI.

One outcome of this collaboration is the recognition of discrete and reoccurring mixtures of CECs that correlate loosely with land use. For example, tributaries in agricultural settings contain more estrogenic compounds, herbicides, and surfactants, whereas urban areas are likely to contain a mixture of prescription pharmaceuticals and personal care products (Elliott et al., 2017). When CECs are found in great numbers and high concentrations, fish glucose (blood sugar) levels increase – a sign of stress often accompanied by changes to the liver. Frequently, fish with high glucose and altered liver anatomy contain smaller reproductive organs and are sexually immature (Thomas et al., 2017). Although there is still a long way to go before the effects of CECs on fish populations are fully understood, early findings provide evidence of detrimental effects worth investigating.

Predicting effects and prioritizing contaminants

Determining ecological risks associated with exposures to complex chemical mixtures is challenging. Effects of many CECs are unknown; they can have additive and/or antagonistic effects and/or interact in unexpected ways and thus may be difficult to predict. Furthermore, most mixture modeling and assessment relies on static mixture exposure scenarios that poorly emulate real-world, fluctuating conditions (Altenburger et al., 2013). Measuring effects of the seemingly endless combination of chemicals using whole-animal regulatory toxicology approaches of the twentieth century is a daunting task; it is unachievable because of time and financial limitations.

With the recent development of high-throughput effects-based tools – a twenty-first-century technology – it is possible to quickly and cheaply screen environmental samples and identify biological targets that these samples affect (Schroeder et al., 2016a). These technologies allow for the simultaneous assessment of hundreds (*in vitro* cell-based assays) to tens of thousands of biological end points (*in vivo* evaluation of expression of all genes, and/or all metabolites). This contrasts strikingly with twentieth-century methodologies where evaluation of a handful of end points was an expensive and ambitious undertaking (Lillicrap et al., 2016). These bioeffects-based monitoring tools measure the integrated biological activity of an entire mixture quickly and have been proposed to help solve these challenges (Schroeder et al 2016a). Bioeffects-based assessment of Great Lakes surface water samples indicated that steroid hormone receptors (estrogen, progesterone, and glucocorticoid) and aryl hydrocarbon receptor responses are typically initiated (Schroeder et al., 2016a; Li et al., 2016). Nevertheless, these studies evaluated only a handful of locations, mostly heavily contaminated sites, including AOCs, so it remains unknown whether these results represent all the Great Lakes. Concerted efforts are underway to use these tools to conduct expansive evaluation (Ekman et al., 2013). While these bioeffects-based approaches represent an important improvement in our ability to assess hazards of complex contaminant mixtures, there are two major limitations to these approaches: (1) they typically measure effects at the molecular level, which can be difficult to extrapolate to organismal and population level, and (2) they do not readily provide insight into which chemicals cause observed biological responses. Overcoming these obstacles is essential if the data generated are to be useful to environmental regulators and managers.

The adverse outcome pathway (AOP) framework has been proposed to address the first obstacle of extrapolation of molecular effects (Ankley et al., 2010). The AOP relies less on *in vivo*–derived organismal data and supports integrated chemical risk assessments by using mechanistic data derived from high-throughput bioeffects-based approaches (Ankley et al., 2010). AOPs and associated data and tools are expected to eventually capture and integrate effects on multiple biological targets, and thus predict and capture mixture effects.

Novel approaches may address the second obstacle – identification of chemical drivers of toxicity. Some approaches use chemical occurrence data for research sites and prior knowledge of toxicity to determine whether enough chemical is present to exert biological effects (Schroeder et al., 2016a; Li et al., 2016). Substances with small exposure-to-effect ratios would indicate low priority for further evaluation, while biological targets that can be initiated by chemicals at measurable concentrations would trigger further work. Other approaches mathematically integrate prior knowledge with on-site testing to identify or exclude chemical drivers of toxicity (Schroeder et al., 2016b). Finally, some approaches use agnostic computational methods to identify drivers of toxicity by combining empirical assessment of biological effects with chemical occurrence data. For example, Davis et al. (2016) used partial least squares regression analyses to integrate field-based metabolomics and chemical occurrence data for the Great Lakes;

they identified 49 CECs that could have been drivers of toxicity and excluded 54 percent of CECs measured at the site.

Conclusion

Prior knowledge-based predictions and high-throughput bioeffects screening approaches of the twenty-first century present an excellent opportunity to identify hazards associated with Great Lakes contaminants. The lakes continue to present complex challenges to understanding ecological risks associated with chemical contamination. It is particularly challenging to determine relative risk associated with new contaminants versus legacy contaminants and nonchemical stressors. Nevertheless, the lakes still serve as an incubator of large-scale research that will advance our understanding of occurrence, fate, and effects of CECs in aquatic ecosystems and yield novel approaches to monitoring and managing contaminants.

References

Altenburger, R., Backhaus, T., Boedeker, W., Faust, M., and Scholze, M. (2013). Simplifying complexity: Mixture toxicity assessment in the last 20 years. *Environmental Toxicology and Chemistry*, *32*(8): 1685–1687.

Ankley, G. T., Bennett, R. S., Erickson, R. J., Hoff, D. J., Hornung, M. W., Johnson, R. D., Mount, D. R., Nichols, J. W., Russom, C. L., Schmieder, P. K., and Serrrano, J. A. (2010). Adverse outcome pathways: A conceptual framework to support ecotoxicology research and risk assessment. *Environmental Toxicology and Chemistry*, *29*(3): 730–741.

Baldwin, A. K., Corsi, S. R., De Cicco, L. A., Lenaker, P. L., Lutz, M. A., Sullivan, D. J., and Richards, K. D. (2016). Organic contaminants in Great Lakes tributaries: Prevalence and potential aquatic toxicity. *Science of the Total Environment*, *554*: 42–52.

Barber, L. B., Loyo-Rosales, J. E., Rice, C. P., Minarik, T. A., and Oskouie, A. K. (2015). Endocrine disrupting alkylphenolic chemicals and other contaminants in wastewater treatment plant effluents, urban streams, and fish in the Great Lakes and Upper Mississippi River regions. *Science of the Total Environment*, *517*: 195–206.

Blair, B. D., Crago, J. P., Hedman, C. J., and Klaper, R. D. (2013). Pharmaceuticals and personal care products found in the Great Lakes above concentrations of environmental concern. *Chemosphere*, 93(9): 2116–2123.

Brodin, T., Fick, J., Jonsson, M., and Klaminder, J. (2013). Dilute concentrations of a psychiatric drug alter behavior of fish from natural populations. *Science*, 339(6121): 814–815.

Carlson, D. L., Vault, D. S. D., and Swackhamer, D. L. (2010). On the rate of decline of persistent organic contaminants in lake trout (Salvelinus namaycush) from the Great Lakes, 1970– 2003. *Environmental Science & Technology*, *44*(6): 2004–2010.

Cook, P. M., Robbins, J. A., Endicott, D. D., Lodge, K. B., Guiney, P. D., Walker, M. K., Zabel, E. W., and Peterson, R. E. (2003). Effects of aryl hydrocarbon receptor-mediated early life stage toxicity on lake trout populations in Lake Ontario during the 20th century. *Environmental Science & Technology*, *37*(17): 3864–3877.

Cordle, F., Corneliussen, P., Jelinek, C., Hackley, B., Lehman, R., McLaughlin, J., Rhoden, R., and Shapiro, R. (1978). Final report of the Subcommittee on health effects of Pcbs and Pbbs: Human exposure to polychlorinated biphenyls and polybrominated biphenyls. *Environmental Health Perspectives*, *24*: 157.

Corsi, S. R., Booth, N. L., and Hall, D. W. (2001). Aircraft and runway deicers at General Mitchell International Airport, Milwaukee, Wisconsin, USA. 1: Biochemical oxygen demand and dissolved oxygen in receiving streams. *Environmental Toxicology and Chemistry*, *20*(7): 1474–1482.

Daughton, C. G. (2005). "Emerging" chemicals as pollutants in the environment: A 21st century perspective. *Renewable Resources Journal*, *23*(4): 6.

Davis, J. M., Ekman, D. R., Teng, Q., Ankley, G. T., Berninger, J. P., Cavallin, J. E., Jensen, K. M., Kahl, M .D., Schroeder, A. L., Villeneuve, D. L., and Jorgenson, Z. G. (2016). Linking field-based metabolomics and chemical analyses to prioritize contaminants of emerging concern in the Great Lakes Basin. *Environmental Toxicology and Chemistry*, *35*(10): 2493–2502.

Denham, M., Schell, L. M., Deane, G., Gallo, M. V., Ravenscroft, J., and DeCaprio, A. P. (2005). Relationship of lead, mercury, mirex, dichlorodiphenyldichloroethylene, hexachlorobenzene, and polychlorinated biphenyls to timing of menarche among Akwesasne Mohawk girls. *Pediatrics*, *115*(2): e127–e134.

Ekman, D. R., Ankley, G. T., Blazer, V. S., Collette, T. W., Garcia-Reyero, N., Iwanowicz, L. R., Jorgenson, Z. G., Lee, K. E., Mazik, P. M., Miller, D. H., and Perkins, E. J. (2013). Biological effects – based tools for monitoring impacted surface waters in the Great Lakes: A multiagency program in support of the Great Lakes Restoration Initiative. *Environmental Practice*, *15*(04): 409–426.

Elliott, S. M., Brigham, M. E., Lee, K. E., Banda, J. O., Choy, S. J., Gefell, D. J., Minarik, T. A., Moore, J. N., Jorgenson, Z. G. (2017) Contaminants of Emerging Concern in Tributaries to the Laurentian Great Lakes: I. Patterns of Chemical Occurrence. *PLOS One*.

Gilbertson, M. (1983). Etiology of chick edema disease in herring gulls in the lower Great Lakes. *Chemosphere*, *12*(3): 357–370.

Gilbertson, M. (1997). Advances in forensic toxicology for establishing causality between Great Lakes epizootics and specific persistent toxic chemicals. *Environmental Toxicology and Chemistry*, *16*(9): 1771–1778.

Gilbertson, M., Kubiak, T., Ludwig, J., and Fox, G. (1991). Great lakes embryo mortality, edema, and deformities syndrome (GLEMEDS) in colonial fish-eating birds: Similarity to chick-edema disease. *Journal of Toxicology and Environmental Health, Part A Current Issues*, *33*(4): 455–520.

Great Lakes Restoration Initiative. (2010). *Great Lakes Restoration Initiative Action Plan.* Availableathttps://yosemite.epa.gov/sab/sabproduct.nsf/fedrgstr_activites/750D38466C98D0818525 771A0068B925/$File/glri_actionplan.pdf

Hala, D., Petersen, L. H., Martinovic, D., and Huggett, D. B. (2012). Constraints-based stoichiometric analysis of hypoxic stress on steroidogenesis in fathead minnows, Pimephales promelas. *Journal of Experimental Biology*, *215*(10): 1753–1765.

Hirai, N., Nanba, A., Koshio, M., Kondo, T., Morita, M., and Tatarazako, N. (2006). Feminization of Japanese medaka (Oryzias latipes) exposed to 17β-estradiol: Effect of exposure period on spawning performance in sex-transformed females. *Aquatic Toxicology*, *79*(3): 288–295.

Hornbuckle, K. C., Carlson, D. L., Swackhamer, D. L., Baker, J. E., and Eisenreich, S. J. (2006). Polychlorinated biphenyls in the Great Lakes. In: Hites, R., (Ed.). Handbook of Environmental Chemistry: *Persistent Organic Pollutants in the Great Lakes*. Heidelberg, Berlin: Springer, 13–70.

Kidd, K. A., Blanchfield, P. J., Mills, K. H., Palace, V. P., Evans, R. E., Lazorchak, J. M., and Flick, R. W. (2007). Collapse of a fish population after exposure to a synthetic estrogen. *Proceedings of the National Academy of Sciences*, *104*(21): 8897–8901.

Li, S., Villeneuve, D. L., Berninger, J. P., Blackwell, B. R., Cavallin, J. E., Hughes, M. N., Jensen, K. M., Jorgenson, Z., Kahl, M. D., Schroeder, A. L., and Stevens, K. E. (2016). An integrated approach for identifying priority contaminant in the Great Lakes Basin – Investigations in the Lower Green Bay/Fox River and Milwaukee Estuary areas of concern. *Science of the Total Environment*, *579*: 825–837.

Lillicrap, A., Belanger, S., Burden, N., Pasquier, D. D., Embry, M. R., Halder, M., L. Lampi, M. A., Lee, L., Norberg-King, T., Rattner, B. A., and Schirmer, K. (2016). Alternative approaches to vertebrate ecotoxicity tests in the 21st century: A review of developments over the last 2 decades and current status. *Environmental Toxicology and Chemistry*, *35*(11): 2637–2646.

Martinović, D., Hogarth, W. T., Jones, R. E., and Sorensen, P. W. (2007). Environmental estrogens suppress hormones, behavior, and reproductive fitness in male fathead minnows. *Environmental Toxicology and Chemistry*, *26*(2): 271–278.

Martinović-Weigelt, D., Minarik, T. A., Curran, E. M., Marchuk, J. S., Pazderka, M. J., Smith, E. A., Goldenstein, R. L., Miresse, C. L., Matlon, T. J., Schultz, M. M., and Schoenfuss, H. L. (2013). Environmental estrogens in an urban aquatic ecosystem: I: Spatial and temporal occurrence of estrogenic activity in effluent-dominated systems. *Environment International*, *61*: 127–137.

Marvin, C., Painter, S., Williams, D., Richardson, V., Rossmann, R., and Van Hoof, P. (2004). Spatial and temporal trends in surface water and sediment contamination in the Laurentian Great Lakes. *Environmental Pollution*, *129*(1): 131–144.

McVeety, B. D., and Hites, R. A. (1988). Atmospheric deposition of polycyclic aromatic hydrocarbons to water surfaces: A mass balance approach. *Atmospheric Environment (1967)*, *22*(3): 511–536.

Newman, R. (1972). "Burn On". Lyrics. Sail Away. Reprise Records.

Painter, M. M., Buerkley, M. A., Julius, M. L., Vajda, A. M., Norris, D. O., Barber, L. B., Furlong, E. T., Schultz, M. M., and Schoenfuss, H. L. (2009.) Antidepressants at environmentally relevant concentrations affect predator avoidance behavior of larval fathead minnows (Pimephales promelas). *Environmental Toxicology and Chemistry*, *28*(12): 2677–2684.

Pompeani, D. P., Abbott, M. B., Steinman, B. A., and Bain, D. J. (2013). Lake sediments record prehistoric lead pollution related to early copper production in North America. *Environmental Science & Technology*, *47*(11): 5545–5552.

Sauvé, S., and Desrosiers, M. (2014). A review of what is an emerging contaminant. *Chemistry Central Journal*, 8(1): 1–7.

Schroeder, A. L., Ankley, G. T., Houck, K. A., and Villeneuve, D. L. (2016a). Environmental surveillance and monitoring – The next frontiers for high-throughput toxicology. *Environmental Toxicology and Chemistry*, *35*(3): 513–525.

Schroeder, A. L., Martinović-Weigelt, D., Ankley, G. T., Lee, K. E., Garcia-Reyero, N., Perkins, E. J., Schoenfuss, H .L., and Villeneuve, D. L. (2016b). Prior knowledge-based approach for associating contaminants with biological effects: A case study in the St. Croix River basin, MN, WI, USA. *Environmental Pollution*, *221*: 427–436.

Solla, S. R. D., Weseloh, D. C., Hughes, K. D., and Moore, D. J. (2016). Forty-year decline of organic contaminants in eggs of herring gulls (*Larus argentatus*) from the Great Lakes, 1974 to 2013. *Waterbirds*, *39*(sp1): 166–179.

Spitsbergen, J. M., Walker, M. K., Olson, J. R., and Peterson, R. E. (1991). Pathologic alterations in early life stages of lake trout, Salvelinus namaycush, exposed to 2, 3, 7, 8- tetrachlorodibenzo-p-dioxin as fertilized eggs. *Aquatic Toxicology*, *19*(1): 41–71.

Stalter, D., Magdeburg, A., Weil, M., Knacker, T., and Oehlmann, J. (2010). Toxication or detoxication? In vivo toxicity assessment of ozonation as advanced wastewater treatment with the rainbow trout. *Water Research*, *44*(2): 439–448.

Thomas, L. M., Jorgenson, Z. G., Brigham, M. E., Choy, S. J., Moore, J. N., Banda, J. A., Gefell, D. J., Minarik, T. A., Schoenfuss, H. L. (2017). Contaminants of emerging concern in tributaries to the Laurentian Great Lakes: II. Biological consequences of exposure. *PLOS One*.

Tillitt, D. E., Cook, P. M., Giesy, J. P., Heideman, W., and Peterson, W. E. (2008). Reproductive impairment of Great Lakes lake trout by dioxin-like chemicals. In: Di Giulio, R. T., and Hinton, D. E. (Eds.), *The Toxicology of Fishes*. Boca Raton, FL: CRC Press.

Turney, W. G. (1971). Mercury pollution: Michigan's action program. *Journal (Water Pollution Control Federation)*: 1427–1438.

Tyler, C., Jobling, S., and Sumpter, J. P. (1998). Endocrine disruption in wildlife: A critical review of the evidence. *Critical Reviews in Toxicology*, *28*(4): 319–361.

Vajda, A. M., Barber, L. B., Gray, J. L., Lopez, E. M., Bolden, A. M., Schoenfuss, H. L., and Norris, D. O. (2011). Demasculinization of male fish by wastewater treatment plant effluent. *Aquatic Toxicology*, *103*(3): 213–221.

Venier, M., Salamova, A., and Hites, R. A. (2015). Halogenated flame retardants in the Great Lakes environment. *Accounts of Chemical Research*, *48*(7): 1853–1861.

Winchester, J. W., and Nifong, G. D. (1971). Water pollution in Lake Michigan by trace elements from pollution aerosol fallout. *Water, Air, and Soil Pollution*, *1*(1): 50–64.

Wright, P. J., and Tillitt, D. E. (1999). Embryotoxicity of Great Lakes lake trout extracts to developing rainbow trout. *Aquatic Toxicology*, *47*(2): 77–92.

10 Water quality in the Great Lakes

Interactions between nutrient pollution, invasive species, and climate change

Gaston E. Small

Introduction

On August 2, 2014, Toledo, Ohio, made headlines after issuing an order to its 450,000 residents to avoid drinking tap water. The city's municipal water supply is drawn from Lake Erie, which was experiencing a bloom of toxic cyanobacteria (blue-green algae) known as *Microcystis*. Although water treatment processes are generally successful in removing microcystin toxins, these toxins had been detected in the post-treatment water, prompting the do-not-drink order. As Toledo residents searched for bottled water, their dependence on the water quality of their neighboring lake became evident.

The Great Lakes have tremendous ecological and economic importance, and they are subject to a wide array of human stressors. The lakes make up nearly 20 percent of the liquid freshwater on Earth and provide drinking water for 40 million people, as well as supporting agricultural and industrial uses. The lakes support a $4 billion sport fishing industry and allow for commercial shipping of more than 200 million tons annually (Great Lakes Environmental Research Laboratory [GLERL]).

In a recent survey, experts identified the effects of invasive species and climate change as having the greatest potential impacts on the Great Lakes ecosystem (Smith et al., 2015). This marks a shift from nonpoint source phosphorus pollution and persistent contaminants that have been the central focus of lake management in the past several decades. More than 180 invasive and non-native species have established populations in the Great Lakes, altering food webs and adversely affecting water quality (Great Lakes Restoration Initiative [GLRI]). Climate change has a range of direct and indirect effects on Great Lakes ecosystems. For example, warming surface water in recent years has lengthened the ice-free season, increased evaporation rates, and lowered lake levels. This affects shoreline ecosystems, as well as human infrastructure and the shipping industry (Posey, 2012). Rising water temperatures may also facilitate the spread of some invasive species (Rahel and Olden, 2008). Finally, the frequency and magnitude of storms affect nutrient fluxes into the lakes and the mixing of the lake water column.

Although each Great Lake faces unique management challenges, Lake Superior and Lake Erie represent two ends of the spectrum of anthropogenic stressors, physical conditions, and biological responses that characterize the lakes. Lake Erie is the shallowest, warmest, and most productive, historically supporting a sizeable commercial fishery (Baldwin et al., 2009). It receives high inputs of nutrients from rivers draining agricultural land. Combined with the shallow morphometry of the lake, this creates ideal conditions for toxic cyanobacteria blooms and a "dead zone" in Lake Erie's Central Basin each summer. The establishment of invasive zebra mussels (*Dreissena polymorpha*) and quagga mussels (*Dreissena bugensis*) in the 1990s has dramatically changed water quality and the structure of the lake food web, which may also promote the growth of nuisance algae (Auer et al., 2010). Climate also plays an important role in water quality, as warming temperatures allow for a longer growing season. In addition, changes in storm frequency and intensity determine nutrient loads entering the lake and the mixing of the water within the lake, setting up conditions for algal blooms and hypoxia (dissolved oxygen < 2 milligrams per liter).

In contrast, in relatively pristine Lake Superior, nitrate levels have increased steadily over the past century due to a combination of the lake's 190-year water residence time and very low baseline phosphorus levels. Warmer temperatures in the last decade have already created shifts in the phytoplankton species. A record-setting storm in 2012 caused significant effects on the lake, and storms of this magnitude may be increasingly common in the future.

Statement of the problem

Humans have altered the Great Lakes ecosystems directly through nutrient pollution and introductions of invasive species, and indirectly through climate change. A warming climate has increased the length of the ice-free season, reduced the mixing of water layers during the warmest months, and led to an increased frequency of major storms. The magnitude and timing of the effects from these human-caused changes also depend greatly on the water residence time in each lake, which is determined by the size and shape of the basins (a relic of Pleistocene glaciation) and the spatial configuration of the lakes. This chapter examines how nutrient pollution interacts with hydrology, climate, and species invasions to determine water quality in Lake Erie and Lake Superior, which represent two ends of a continuum of biological, chemical, and physical conditions.

Findings

Lake Erie: a legacy of nutrient pollution, invasion, and toxic algae

Lake hydrology and nutrient status

Although the Great Lakes appear from maps and satellite photographs to be similar in size, these images belie important differences. Lake Erie is much shallower than the others and is divided into three distinct basins: the shallow Western Basin (average depth 7 meters [23 feet]), the Central Basin (average depth 18

Table 10.1 Physical and hydrologic characteristics of the Great Lakes. Lake watershed area represents only the area draining directly into that lake. Total basin area includes the watersheds of upstream lakes.

	Lake Superior	*Lake Michigan*	*Lake Huron*	*Lake Erie*	*Lake Ontario*
Lake area (km^2)	82,103	58,026	59,588	25,847	18,960
Average depth (m)	147	85	59	7.4 (Western Basin) 18.3 (Central Basin) 64 (Eastern Basin)	86
Lake volume (km^3)	12,100	4,918	3,540	483	1,639
Watershed area (km^2)	127,700	118,000	134,400	78,000	64,000
Total basin area (km^2)	127,700	118,000	380,100	458,100	522,100
Basin area: Lake volume (km^2/km^3)	10.5	23.0	107.4	948.4	318.5
Water residence time (years)	191	99	22	2.6	6

meters [59 feet]), and the deepest Eastern Basin (average depth 64 meters [210 feet]). Because of its shallow depth relative to the other Great Lakes, it has a much smaller volume (it is 1/25th the size of Lake Superior). Additionally, it is downstream of lakes Superior, Michigan, and Huron, so the total basin size contributing water to Lake Erie is larger than its three upstream neighbors. The influence of a watershed on a lake can be described by the ratio of the drainage basin area to the volume of a lake. Every cubic kilometer of water in Lake Erie corresponds to 948 square kilometers of watershed (1,525 square miles of watershed per cubic mile of water), much larger than any of the other lakes (Table 10.1). For comparison, Lake Superior has a ratio of 10.5 square kilometers of watershed for every cubic kilometer of lake water (17 square miles of watershed per cubic mile of water). This ratio of drainage basin area to lake volume also influences the residence time of water in the lake. Water in Lake Erie turns over every 2.6 years, compared to 191 years for Lake Superior. Lake Erie is more heavily influenced by agriculture and urbanization than the upstream lakes surrounded by large cities (12 million people live in the watershed, which includes 17 metropolitan areas) and receives water and nutrients from the Maumee and Sandusky rivers, where cultivated agriculture is a major land use (Ohio EPA).

Although concentrations of major ions have increased in all of the Great Lakes since the mid-1800s due to human influence, Lake Erie in particular experienced the greatest changes in water quality and biology (Beeton, 1965). During the 1960s, Lake Erie was perceived to be "dying" due to excessive algal blooms, and by the late 1960s regulatory agencies agreed that high levels of phosphorus (P) inputs were primarily responsible (Box 10.1).

Box 10.1 Quantifying water quality

Water quality is typically quantified in terms of chlorophyll concentration, an indicator of phytoplankton biomass. Phosphorus (P) is often a limiting nutrient for algal growth in freshwater, and total P (the P in dissolved and particulate forms in a water sample) generally correlates with chlorophyll levels. Likewise, reductions in P inputs are typically required to significantly improve lake water quality. Because phytoplankton densities typically determine lake water clarity, measurements of turbidity such as Secchi disk depths (the depth at which a sharply patterned disk disappears from view upon lowering from a boat) are also often used as metrics of water quality. Clear lakes with chlorophyll *a* levels less than 2.6 micrograms/liter, total P concentrations below 12 micrograms/liter, and Secchi depths greater than 4 meters (13 feet), are considered oligotrophic. Lakes with chlorophyll *a* values from 20–56 micrograms/liter, total P concentrations from 24–96 micrograms/liter, and Secchi depths 0.5–2 meters (1.6–6.5 feet), are considered to be eutrophic. Hypereutrophic lakes contain higher chlorophyll and total P concentrations, and mesotrophic lakes have values between oligotrophic and eutrophic ranges (Carlson and Simpson, 1996).

Phytoplankton produces oxygen through photosynthesis as it grows, but after it dies and sinks, decomposition consumes oxygen. In shallow lakes, mixing by wind keeps the water column oxygenated, but deeper lakes are often thermally stratified during the summer, with the deeper layer separated from the atmosphere. Oxygen consumption by decomposing phytoplankton can lead to hypoxia in the hypolimnion, making this region of the lake uninhabitable by organisms that require oxygen – potentially leading to fish die-offs.

The first Great Lakes Water Quality Agreement, signed in 1972, aimed to reduce P inputs by more than half, from 29,000 to 14,600 metric tons per year (U.S. EPA). The United States and Canada agreed to further reductions in P pollution in 1978, to 11,000 metric tons per year. These reductions were achieved by eliminating P in laundry detergent, upgrading wastewater treatment plants, and implementing best management practices for agricultural lands. These reductions in P inputs had the intended effect in the most polluted Western Basin: from the late 1960s to the mid-1980s, there was a fivefold decrease in phytoplankton density in western Lake Erie (Nicholls and Hopkins, 1993). Changes were less pronounced in the Central and Eastern basins, although the rate of seasonal oxygen depletion in the Central Basin (i.e. the onset of hypoxia) did decrease slightly during the 1980s (Bertram, 1993).

Effects of invasive dreissenid mussels

Invasion by zebra mussels in the late 1980s, followed shortly by the closely related quagga mussels, led to further dramatic reductions in phytoplankton density due to these filter-feeding bivalves. Two years after zebra mussels were established in Lake Erie, chlorophyll *a* concentrations (a standard measurement of phytoplankton abundance) in the lake were < 1 microgram/liter for most of 1991, typical of the oligotrophic upper Great Lakes (Nicholls and Hopkins, 1993). Although zebra mussels led to significant decreases in chlorophyll and water column total P, and increases in Secchi depth in Lake Erie and other areas (e.g., Saginaw Bay in Lake Huron), researchers argued that these changes did not alter the trophic status of the lakes because excess nutrients are still part of the lake ecosystem. Rather, these organisms have changed the spatial partitioning of resources, moving phosphorus from the water column to the sediment (Fahnenstiel et al., 1995). The density of *Dreissena* mussels peaked in the Central and Eastern basins from 1998 to 2002, and has since declined (Karatayev et al., 2014). Quagga mussels are now dominant in these basins, while zebra mussels coexist with quagga mussels in the shallow Western Basin.

An example of the complex ecological effects of zebra mussels is the increase in growth of the nuisance algae *Cladophora* in Lake Erie, as well as in lakes Michigan and Ontario. It grows attached to rocks in shallow waters, and shorelines can be covered with piles of decaying algae. Reduced P inputs through the mid-1980s were sufficient to suppress *Cladophora* growth, but, following zebra mussel establishment and associated increases in water clarity, it was able to grow in deeper waters, and therefore total production of the nuisance algae gradually increased (Auer et al., 2010).

The re-eutrophication of Lake Erie

Although P loading rates have generally been at or below the target of 11,000 metric tons per year since 1980 (Dolan and Chapra, 2012), toxic and nuisance algae blooms in the lake have increased since the 1990s, and bottom-water hypoxia has been a seasonal occurrence (Scavia et al., 2014). This decline in water quality may be due to a several factors. Warmer water temperatures can increase algal growth as a result of longer growing seasons. More frequent severe storms during spring can lead to pulses of nutrients moving from urban and agricultural land into the lake. While total P input has stayed relatively constant, the fraction of this P that enters in dissolved form (readily accessible to phytoplankton) has increased in recent years (U.S. EPA). The shallow Western Basin receives 61 percent of the lake's P inputs, and these waters experience regularly occurring blooms by the cyanobacteria *Microcystis aeruginosa* during spring and summer. Reducing the area of the hypoxic zone to the levels observed in the 1990s would require an estimated 46 percent cut in total P inputs to the lake, or a 78 percent reduction in dissolved P loads (Scavia et al., 2014).

Local weather conditions play an important role in mediating the effects of nutrient loading. As a result, changes in storm frequency due to climate change could have large effects on water quality. During the summer of 2011, for example, western Lake Erie experienced a record-setting *Microcystis* bloom (Figure 10.1) with concentrations three times higher than previously observed. During this bloom, microcystin toxin concentrations in the water were 50 times greater than World Health Organization limits for safe body contact, and 1,200 times higher than safe drinking water limits. The intensity of the bloom was attributed to high P inputs from agriculture to the Western Basin and an extended period of calm weather that led to long water residence times and a stable water column, allowing the algae to incubate undisturbed at the surface (Michalak et al., 2013).

Figure 10.1 Harmful algae bloom in Lake Erie, July 2011.

Source: NOAA Great Lakes Environmental Research Laboratory

This combination of factors is expected to occur more frequently due to climate change, likely leading to increasingly frequent large harmful algal blooms. In addition to facilitating these blooms, lower storm frequency in the summer leads to a stable water column, amplifying the effects of nutrient inputs on hypoxia in the Central Basin (Conroy, 2008).

Nitrogen: an underappreciated contributor to Lake Erie water quality

Although phosphorus has been recognized as the key nutrient to controlling algal growth in lakes since the early 1970s, the availability of dissolved nitrogen (N) is an important determinant of the type of phytoplankton present. Green algae and diatoms are generally abundant when N and P are both available, but low levels of N favor the growth of varieties of cyanobacteria (blue-green algae) able to "fix" atmospheric nitrogen gas (N_2).

Algal growth in Lake Erie is limited by a combination of elements including N (North et al., 2007). Lake Erie receives higher inputs of N from rivers that drain agricultural and urban areas, and, to a lesser extent, from the upstream lakes than the other Great Lakes (GLEAM 2016), yet has the lowest dissolved N concentrations of the lakes (U.S. EPA 2016), because of the active biological processing of this element. By summer, most available N has already been sequestered into phytoplankton or removed from the lake through microbial denitrification. In that process, bacteria in anoxic environments consume nitrate in place of oxygen, respiring inert N_2 gas as a waste product. Denitrification rates are high in Lake Erie as the high amounts of organic carbon in the sediment from decomposing algae mean that oxygen in the sediment disappears within millimeters of the sediment-water interface, creating ideal conditions for denitrification. Lake Erie has the highest denitrification rates among the Great Lakes (Small et al., 2013; Small et al., 2016), which increase in response to additional nitrate availability, effectively stabilizing the lake's low nitrate levels.

In July 2011 during the record-setting *Microcystis* bloom in Lake Erie (*Microcystis* is a non-N-fixing cyanobacteria), the author and a team happened to be conducting an 18-day research cruise studying N cycling across lakes Superior, Huron, and Erie. Longitudinal/depth profiles from this transect illustrate how different conditions in Lake Erie compare to these two upstream lakes. Lake Erie surface water was 25°C (77°F), and oxygen depletion was becoming evident in the Central Basin hypolimnion. Chlorophyll *a* concentrations in the Western Basin were three- to fourfold higher compared to lakes Superior and Huron. Nitrate concentrations remained high (approximately 420 micrograms N/L as NO_3^-) in the Western Basin but were diminished (below 140 micrograms N/L as NO_3^-) throughout the Central and Eastern basins. By contrast, ammonium – the reduced form of dissolved N – present in low levels throughout lakes Superior and Huron, was high in the deep waters of Lake Erie's Eastern Basin, indicating high rates of decomposition of organic matter in the sediment. Higher concentrations of nitrogen gas indicate denitrification;

dissolved nitrogen was elevated in the Western Basin, while N_2O – the product of incomplete denitrification – was elevated in the Central Basin hypolimnion.

Our analysis of Lake Erie's N budget indicated that approximately half (48±7 percent) of N that enters Lake Erie ultimately gets removed by denitrification within the lake while the remainder ultimately is transported downstream to Lake Ontario. Nitrogen atoms are transformed on average 15 times during their lifetime in Lake Erie. Of the N removed from denitrification in Lake Erie, approximately 75 percent originated as direct inputs from tributaries to Lake Erie, 21 percent originated as inputs to Lake Huron, and 4 percent originated in Lake Superior (Small et al., 2014). The active processing and removal of N in Lake Superior adds a dynamic component to the water quality story in Lake Erie, and is both a cause (e.g., denitrification leading to conditions favoring blooms of N-fixing cyanobacteria) and an effect (e.g., hypoxia driven by decomposing phytoplankton, which creates conditions favoring denitrification) of the lake's periodic algal blooms.

Lake Superior: slow but steady changes in Earth's largest freshwater lake

Nitrate buildup in Lake Superior: a pristine paradox

Lake Superior is, by any definition, oligotrophic. Chlorophyll *a* levels are low (~ 0.5 micrograms/liter), and Secchi depths extend approximately 10 meters (33 feet). Phosphorus levels are so low that they are near the limit of analytical detection. Positioned at the upstream terminus of the chain of Great Lakes, Lake Superior has a relatively small watershed that is predominantly forested and sparsely populated (Table 10.1).

It may seem a paradox, then, that nitrate levels in the lake have increased more than fivefold in the past century, from ~70 micrograms N/liter as NO_3^- in 1905 to more than 350 micrograms N/liter as NO_3^- today (Sterner et al., 2007; Great Lakes Monitoring). There are now 10,000 N atoms (in the form of nitrate) for every one P atom (in the form of phosphate) in Lake Superior water (Sterner et al., 2007), a striking deviation from the physiological requirements for algal growth (an N:P ratio of approximately 16:1).

The amount of nitrate buildup in the lake each year is approximately equal to the amount of nitrate that flows downstream into Lake Huron (Sterner et al., 2007). Isotopic evidence indicates that most nitrate in the lake has been transformed from other forms of N within the lake (Finlay et al., 2007), indicating active N cycling. Our measurements of nitrification rates in Lake Superior found that rates were relatively low – two orders of magnitude lower than most lakes, but similar to rates reported for the coastal ocean (Small et al., 2013a). However, Lake Superior's nitrification rate is still more than 50 times higher than the rate of long-term nitrate buildup in the lake, further evidence that N cycles repeatedly once it enters the lake. Our analysis of N cycling indicated that the average N atom that enters changes approximately 47 times before exiting the lake either through denitrification or moving downstream. That compares to 26 average transformations in Lake Huron and 15 in Lake Erie (Small et al., 2014). Although the actual biogeochemical rates in Lake Superior are relatively low in the amount of nutrient transformed per liter of water per hour, the very long residence time of water in Lake Superior allows for many of these transformations during the lifespan of nutrients in the lake.

The ultimate causes of this increase are not entirely understood; its 191-year residence time of water means that a slight change in N inputs could have long-term effects. The combustion of fossil fuels puts reactive nitrogen in the atmosphere, which returns through precipitation or dry deposition. The resulting nitrogen inputs from atmospheric deposition have increased in the past century. It is also possible that logging in Lake Superior's watershed in the early twentieth century, followed by a shift in forest composition from hardwoods to softwoods, could have led to slight increases in the amount of nitrogen entering the lake. Changes in other conditions within the lake could also contribute to the nitrate rise.

Recent research has improved our understanding of the conditions in the lake that have allowed for this nitrate rise. Because of relatively low amounts of organic carbon in the sediment, oxygen penetrates several centimeters into the lake sediment – approximately 10 times deeper than in Lake Erie. Denitrification requires anoxic conditions to remove nitrate, meaning nitrate in the lake water must diffuse through this thicker layer of sediment to reach the active denitrification zone. Moreover, nitrate is also being produced from nitrification within the oxygenated sediment layer. From sediment core incubations, we found Lake Superior sediment is a net source of nitrate to the water column, and denitrification in the anoxic sediment depths removes only nitrate produced from within the sediment, not from the water column (Small et al., 2013b). Moreover, this low rate of denitrification is already maxed out. In contrast to Lake Erie, where potential denitrification rates jump up if you add more nitrate, Lake Superior already has abundant nitrate, and adding more has no effect on rates (Small et al., 2016). This means there is no built-in "buffering capacity" for stabilizing nitrate levels in Lake Superior. The slightest increase in N inputs result in nitrate levels building up. That is because they will not be balanced by increased phytoplankton uptake that is limited by P, or by denitrification.

These low denitrification rates seem to be the rule for offshore Lake Superior. It should be noted that in the shallow, near-shore waters of western Lake Superior near Duluth, Minnesota, and in some protected inshore waters – the Saint Louis River Estuary on the lake's western margin and parts of Nipigon and Black bays on the north side – we measured denitrification rates similar to those in Lake Erie (Small et al., 2016). However, these areas are small compared to the total lake area.

This nitrate buildup in Lake Superior is not unique – a similar buildup has been observed in other large lakes worldwide, including Lake Ontario, and to a lesser extent, lakes Huron and Michigan. As P inputs to these lakes have been reduced in recent decades due to stricter regulations, decreased phytoplankton growth results in less organic matter entering the lake, so less N is lost through permanent burial or denitrification (Finlay et al., 2013). Paradoxically, improving lake water quality in terms of the traditional metrics of chlorophyll *a*, total P, and water clarity has the unintended side effect of reducing rates of N removal, thus resulting in nitrate buildup.

There is yet another paradox. Despite the low denitrification rates and increasing nitrate levels in Lake Superior, it is still more efficient at removing N than other Great Lakes. Lake Superior ultimately removes an estimated 84 percent of N that enters the lake through denitrification or permanent burial, contrasted with 16 percent that ultimately moves downstream into Lake Huron. Lake Huron removes an estimated 64 percent of N entering the lake from denitrification, and Lake Erie removes only 48 percent (Small et al., 2014). Again, this is due to the many opportunities each N atom has for denitrification during its long time in the lake. Even though water column N is not directly removed through denitrification, some N atoms are still taken up by phytoplankton, sinking and becoming part of the sediment. Upon decomposition, this organic N is converted into ammonia, which is then transformed to nitrate, and finally a fraction of this nitrate is removed through denitrification. Even though most N in Lake Superior will ultimately be removed through denitrification, nitrate continues to build up because it is generated slightly faster than it is removed. Given the inertia of Earth's largest lake, this buildup will almost certainly continue until a new equilibrium is reached, either from increased denitrification rates – unlikely unless P loading increases – or from increased N export downstream.

Effects of invasive species and climate on water quality in Lake Superior

Ninety-seven non-native species are known to live in Lake Superior (Minnesota Sea Grant), including 19 fish species, 31 species of aquatic invertebrate, 14 parasites, 7 algae species, 3 aquatic vascular plants, and 20 wetland plants, but these species appear to have had a limited impact on water quality so far. Zebra and quagga mussels have established populations only in the harbors at Duluth and Thunder Bay, Ontario (GLEAM); they are likely unable to colonize in the cold, deep water that characterizes most of the lake. The large volume of lake water, combined with the low levels of P and phytoplankton, may minimize the impact on water quality from species invasions.

Climate change is also altering biotic communities in Lake Superior. Since 2001, it has seen an increase in small-celled cyanobacteria during the spring and a decrease in diatoms during the summer, which may be due to lake warming (Reavie et al., 2014). Warming is already changing competitive interactions among fish species in Lake Superior. Between 1979 and 2006, the spatial extent of preferred thermal habitat has increased by 400–500 square kilometers per year (156–193 square miles per year) for lean lake trout (also known as "common lake trout"), Chinook salmon, and walleye, while siscowet lake trout have lost optimal habitat at a rate of more than 150 square kilometers per year (58 square miles per year) (Cline et al., 2013). These habitat alterations could lead to changes in the lake food web, and may make conditions more suitable for some invasive species. Whether these biotic changes will lead to changes in lake water chemistry remains to be seen.

Changes in storm frequency and magnitude may also affect Lake Superior water quality. On June 19–20, 2012, a "mega-rain event" produced up to 25 centimeters (10 inches) of rain along the western Lake Superior watersheds. Discharge from the Saint Louis River reached its highest levels in its more than 100-year recorded history (Czuba et al., 2012), and Lake Superior's water level rose 8–10 centimeters (3–4 inches). A large plume along the southwest margin of the lake was evident from satellite photos for the next week. Shipboard sampling showed the plume to have highly elevated concentrations of suspended sediments, total P, and dissolved P (Minor et al., 2014). Based on the surface area of the plume measured from satellite images and the plume depth and P concentrations that we measured on a research cruise five days after the flood, the plume from this mega-rain event contributed approximately 600,000 kilograms (661 tons) of P to Lake Superior. That corresponds to 40 percent of the total P annual inputs to the lake for an average year. Despite this massive input of P to a P-limited lake ecosystem, increases in chlorophyll were not detected at nearshore monitoring sites in the weeks and months following the flood, presumably because the plume also inhibited light availability for phytoplankton (Minor et al., 2014). However, this P would still ultimately be available to fuel production in the lake. Indeed, increased levels of cyanobacteria were observed near the plume along the Apostle Islands National Lakeshore (Larson et al., 2012).

Conclusions

Water quality in the Great Lakes is controlled by a combination of factors, including physical characteristics such as water residence time, natural and anthropogenic inputs of P and N, and the biotic community's ability to transform these nutrients into biomass. Climate change and species invasions have been and will continue to affect the Great Lakes, but how these effects play out will vary lake-to-lake. Lakes Erie and Superior represent ends of a gradient of physical and biological conditions, and understanding controls on water quality in these lakes will allow us to better anticipate future changes in the Great Lakes.

Box 10.2 Invasive species in the Great Lakes

The Great Lakes have been altered by the invasion and establishment of more than 180 non-native species. Most of these invasions have occurred since the 1959 opening of the St. Lawrence Seaway, a system of canals and locks that allows oceangoing ships throughout the Great Lakes. Many of these species are native to the Black and Caspian seas, including the zebra mussel and related quagga mussel; the fishhook waterflea; and fish species such as the ruffe, tubenose goby, and Blueback herring (Ricciardi and MacIsaac, 2000).

There is some evidence that as more species from the Black and Caspian seas become established in the Great Lakes, the food web of their native ecosystem is being reassembled. This in turn facilitates establishment of additional invaders from this region. This cascade is an "invasion meltdown" (Ricciardi and MacIsaac, 2000).

References

Auer, M. T., Tomlinson, L. M., Higgins, S. N., Malkin, S. Y., Howell, E. T., and Bootsma, H. A. (2010). Great Lakes Cladophora in the 21st century: Same algae – different ecosystem. *Journal of Great Lakes Research*, *36*: 248–255.

Baldwin, N. S., Saalfeld, R. W., Dochoda, M. R., Buettner, H. J., and Eshenroder, R. L. (2009). *Commercial Fish Production in the Great Lakes 1867–2006*. Great Lakes Fishery Commission, Ann Arbor, Michigan.

Beeton, A. M. (1965). Eutrophication of the St. Lawrence Great Lakes. *Limnology and Oceanography*, *10*: 240–254.

Bertram, P. E. (1993). Total phosphorus and dissolved oxygen trends in the Central Basin of Lake Erie, 1970–1991. *Journal of Great Lakes Research*, *19*: 224–236.

Carlson, R. E., and Simpson, J. (1996). A coordinator's guide to volunteer lake monitoring methods. *North American Lake Management Society*, 96: 305.

Cline, T. J., Bennington, V., and Kitchell, J. F. (2013). Climate change expands the spatial extent and duration of preferred thermal habitat for Lake Superior fishes. *PLOS One*, 8(4): e62279.

Conroy, J. D. (2008). Comparing annual trends of dissolved oxygen depletion in Lake Erie: The importance of weather. *Ohio Journal of Science*, *108*: A8–A9.

Czuba, C. R., Fallon, J. D., and Kessler, E. W. (2012). *Floods of June 2012 in Northeastern Minnesota*. U.S. Geological survey Scientific Investigations Report 2012–5283.

Dolan, D. M., and Chapra, S. C. (2012). Great Lakes total phosphorus revisited: 1: Loading analysis and update (1994–2008). *Journal of Great Lakes Research*, 38: 730–740.

Fahnenstiel, G. L., Lang, G. A., Nalepa, T. F., and Johengen, T. H. (1995). Effects of Zebra Mussel (Dreissena polymorpha) colonization on water quality parameters in Saginaw Bay, Lake Huron. *Journal of Great Lakes Research*, *21*: 435–448.

Finlay, J. C., Small, G. E., and Sterner, R. W. (2013). Human influences on nitrogen removal in lakes. *Science*, *342*: 247–250.

Finlay, J. C., Sterner, R. W., and Kumar, S. (2007). Isotopic evidence for in-lake production of accumulating nitrate in Lake Superior. *Ecological Applications*, 17: 2323–2332.

Great Lakes Environmental Assessment and Mapping Project (GLEAM). (2016). Available at www.greatlakesmapping.org/great_lake_stressors/7/nitrogen-loading

Great Lakes Environmental Research Laboratory (GLERL). (2016).*National Oceanic and Atmospheric Administration*. Available at www.glerl.noaa.gov/pr/ourlakes/intro.html

Great Lakes Monitoring. (2016). *Illinois-Indiana Sea Grant*. Available at https://greatlakes-monitoring.org/geodashboard/

Great Lakes Restoration Initiative (GLRI). (2016). *National Oceanic and Atmospheric Administration*. Available at www.regions.noaa.gov/great-lakes/index.php/great_lakes-restoration-initiative/invasive-species/

Karatayev, A. Y., Burlakova, L. W., Pennuto, C., Ciborowski, J., Karatayev, V. A., Juette, P., and Clapsadl, M. (2014). Twenty five years of changes in Dreissena spp: Populations in Lake Erie. *Journal of Great Lakes Research*, *40*: 550–559.

Larson, N. J., LaLiberte, G. D., and Van Stappen, J. (2012, July 23). *Blue-green Algae Observed in Lake Superior*. Joint press release – Wisconsin Department of Natural Resources and Apostle Islands National Seashore.

Michalak, A. M., Anderson, E. J., Beletsky, D., Boland, S., Bosch, N. S., Bridgeman, T. P., Chaffin, J. D., Cho, K., Confessor, R., Daloğlu, I., DePinto, J. V., Evans, M. A., Fahnenstiel, G. L., He, L., Ho, J. C., Jenkins, L., Hohengen, T. H., Kuo, K. C., LaPorte, E., Liu, X., McWilliams, M. R., Moore, M. R., Posselt, D. J., Richards, R. P., Scavia, D., Steiner, A. L., Verhamme, E., Wright, D. M., and Zagorski, M. A. (2013). Record-setting algal bloom in Lake Erie caused by agricultural and meteorological trends consistent with expected future conditions. *Proceedings of the National Academy of Sciences, U.S.A.*, *110*: 6448–6452.

Minnesota Sea Grant. (2016). *Lake Superior's nonnative fish species*. Available at www.seagrant.umn.edu/ais/superior_nonnatives

Minor, E. C., Forsman, B., and Guilfond, S. J. (2014). The effect of a flood pulse on the water column of western Lake Superior, USA. *Journal of Great Lakes Research*, *40*: 455–462.

Nicholls, K. H., and Hopkins, G. J. (1993). Recent changes in Lake Erie (North Shore) phytoplankton: Cumulative impacts of phosphorus loading reductions and the Zebra Mussel introduction. *Journal of Great Lakes Research*, *19*: 637–647.

North, R. L., Guidford, S. J., Smith, R. E. H., Havens, S. M., and Twiss, M. R. (2007). Evidence for phosphorus, nitrogen, and iron colimitation of phytoplankton communities in Lake Erie. *Limnology and Oceanography*, *52*: 315–328.

Ohio Environmental Protection Agency (Ohio EPA). (2016). *Maumee River Watershed*. Available at www.epa.state.oh.us/dsw/tmdl/MaumeeRiver.aspx

Posey, J. (2012). *Climate Change Impacts on Transportation in the Midwest*. U.S. National Climate Assessment, Midwest Technical Input Report.

Rahel, F. J., and Olden, J. D. (2008). Assessing the effects of climate change on aquatic invasive species. *Conservation Biology*, *22*: 521–533.

Reavie, E. D., Barbiero, R. P., Allinger, L. E., and Warren, G. J. (2014). Phytoplankton trends in the Great Lakes, 2001–2011. *Journal of Great Lakes Research*, *40*: 618–639.

Ricciardi, A., and MacIsaac, H. J. (2000). Recent mass invasion of the North American Great Lakes by Pont-Caspian species. *Trends in Ecology & Evolution*, *15*: 62–65.

Scavia, D., Allan, J. D., Arend, K. K., Bartell, S., Beletsky, D., Bosch, N. S., Brandt, S. B., Briland, R. D., Daloğu, I., DePinto, J. V., Dolan, D. M., Evans, M. A., Farmer, T. M., Goto, D., Han, H., Höök, T. O., Knight, R., Ludsin, S. A., Mason, D., Michalak, A. M., Richards, R. P., Roverts, J. J., Rucinski, D. K., Rutherford, E., Schwab, D. J., Sesterhenn, T. M., Zhang, H., and Zhou, Y. (2014). Assessing and addressing the re-eutrophication of Lake Erie: Central basin hypoxia. *Journal of Great Lakes Research*, *40*: 226–246.

Small, G. E., Bullerjahn, G. S., Sterner, R. W., Beall, B. F. N., Brovold, S., Finlay, J. C., McKay, R. M. L., and Mukherjee, M. (2013a). Rates and controls of nitrification in a large oligotrophic lake. *Limnology & Oceanography*, *58*: 276–286.

Small, G. E., Cotner, J. B., Finlay, J. C., Stark, R. A., and Sterner, R. W. (2013b). Nitrogen transformations at the sediment-water interface across redox gradients on the Laurentian Great Lakes. *Hydrobiologia*, *731*: 95–108.

Small, G. E., Finlay, J. C., McKay, R. M. L., Rozmarynowycz, M. J., Brovold, S., Bullerjahn, G. S., Spokas, K., and Sterner, R. W. (2016). Large differences in potential denitrification and sediment microbial communities across the Laurentian Great Lakes. *Biogeochemistry*, *128*: 353–368.

Small, G. E., Sterner, R. W., and Finlay, J. C. (2014). An Ecological Network Analysis of nitrogen cycling in the Laurentian Great Lakes. *Ecological Modelling*, *293*: 150–160.

Smith, S. D. P., McIntyre, P. B., Halpern, B. S., Cooke, R. M., Marino, A. L., Boyer, G. L., Buchsbaum, A., Burton Jr., G. A., Campbell, L. M., Ciborowski, J. J. H., Doran, P. J., Infante, D. M., Johnson, L. B., Read, J. G., Rose, J. B., Rutherford, E. S., Steinman, A. D., and Allan, J. D. (2015). Rating impacts in a multi-stressor world: A quantitative assessment of 50 stressors affecting the Great Lakes. *Ecological Applications*, *25*: 717–728.

Sterner, R. W., Anagnostou, E., Brovold, S., Bullerjahn, G. S., Finlay, J. C., Kumar, S., McKay, R. M. L., and Sherrell, R. M. (2007). Increasing stoichiometric imbalance in North America's largest lake: Nitrification in Lake Superior. *Geophysical Research Letters*, *34*. doi: 10.1029/2006GL028861

U.S. Environmental Protection Agency (US EPA). (2016). *Recommended Binational Phosphorus Targets*. Available at www.epa.gov/glwqa/recommended-binational-phosphorus-targets

11 Emerald ash borer, black ash, and Native American basketmaking

Invasive insects, forest ecosystems, and cultural practices

Therese M. Poland, Marla R. Emery, Tina Ciaramitaro, Ed Pigeon, and *Angie Pigeon*

Introduction

The significance of forest resources to Native American culture

Native cultures coevolved with the forests of the Great Lakes region following the last ice age. Plentiful water, abundant game, and fertile soil supported fishing, hunting, and gathering, as well as subsistence agriculture. Lakes and tributaries facilitated transportation by canoe and trade among tribes. Native Americans developed a semi-nomadic lifestyle, moving seasonally among camps as they harvested and cultivated foods, medicines, supplies, and ceremonial items (Kurtz et al., 2015). They relied on natural resources for clothing, shelter, and food. Their cultures continue to rely on natural resources, including trees and understory plants (wood, bark, branches, leaves, and nuts), nonvascular plants, fungi, and animals.

Many species have special cultural and historical value to tribes. In the Great Lakes region, culturally significant species include sugar maple (*Acer saccharum*), paper birch (*Betula papyrifera*), northern white cedar (*Thuja occidentalis*), hop hornbeam (*Ostrya virginiana*), balsam fir (*Abies balsamea*), and black ash (*Fraxinus nigra*). One of the best-known uses of sugar maple is for syrup and sugar from sap, which is so culturally significant that the Anishinaabe people named one of the 13 moons of the year as the "maple sugar moon" (Erickson, 2006). It is also used for traditional medicine (Meeker et al., 1993). With proper techniques, the bark of the paper birch may be removed from live trees without harm. Bark characteristics vary and, depending on the particular tree, may be suitable for different uses (Emery et al., 2014) including canoes, waterproof shelters, baskets, buckets, and food containers (Morgenstern, 2006).

Tribes have many cultural uses for northern white cedar, which holds great spiritual meaning, and it plays a central role in traditional sacred stories and is honored with the name *Nookomis Giizhik*, "Grandmother Cedar" (Densmore, 1974). The wood is used for canoe push poles, ricing knockers for wild rice, spigots for tapping and paddles for stirring maple syrup, cradle boards, and canoe ribs and planks. The bark is used to cover lodges, as an insect repellant, and for bedding. Leaves, twigs, and buds are used for incense, tea, wreaths, decorations, and medicine (Danielsen, 2002). Hop hornbeam bark and inner wood are used for medicinal purposes. In addition, its buds and catkins are an important food source for wild turkey, sharp-tailed grouse, and other wildlife significant to Native American culture (Ritz, 2012). Tribes use balsam fir needles for medicine and its sap or resin blisters to caulk boats. It also provides food for moose, American red squirrels, crossbills, and chickadees, as well as shelter for moose, snowshoe hares, white-tailed deer, ruffed grouse, and other small mammals and songbirds important to Native American culture (Fuller, 2015).

Black ash has special importance for American Indian and First Nations peoples in the Great Lakes region. Its ring-porous wood allows layers of xylem to be easily separated (U.S. Department of Agriculture Natural Resources Conservation Service, 2014) into strips for basket-making. Its wood is used for fish weirs, pipe stems, and lacrosse sticks (Benedict and David, 2003; Mundell, 2008). Revival of the centuries-old tradition of black ash basketmaking is part of the cultural renaissance among tribes in the region and central to the household economies of skilled Native American artisans.

Invasive species threaten culturally important tree species

Invasive forest insects threaten these culturally important species. For example, the Asian long-horned beetle, *Anoplophora glabripennis* (Coleoptera: Cerambycidae), imperils urban and forest hardwoods and has a broad host range but prefers maples in North America (Haack et al., 2006). The wood-boring Japanese cedar long-horned beetle, *Callidiellum rufipenne* (Coleoptera: Cerambycidae), attacks only weakened or freshly felled conifers in its native East Asian range. Eastern U.S. infestations have been found in healthy American arborvitae (northern white cedar) nursery stock and reared from logs of northern white cedar, *Juniperus virginiana* (eastern red cedar), and *Chamaecyparis nootkatensis* (Alaska or yellow cedar) (Maier, 2007). The balsam woolly adelgid, *Adelges piceae* (Hemiptera: Adelgidae), is a tiny sucking insect native to Eurasia where host trees are relatively resistant, but it is a significant invasive pest in both eastern and western North America where it has caused significant mortality to true firs. Infestations occur on balsam fir in the Northeast (Ragenowich and Mitchell 2006).

The emerald ash borer *Agrilus planipennis* Fairmaire (Coleoptera: Buprestidae) is the most destructive invasive forest insect yet to infest North America. It threatens North America's ash resources, including black ash (Poland and McCullough, 2006) and the Indigenous cultures and traditions that rely on it.

Statement of the issue

The challenge of responding to invasive species such as the emerald ash borer while preserving Indigenous cultural traditions requires integrating information from both traditional ecological knowledge and Western science. Traditional ecological knowledge is a "cumulative body of knowledge, practice, and belief, evolving by adaptive processes and handed down through generations by cultural transmission, about the relationship of living beings (including humans) with one another and with their environment" (Berkes, 1999). It provides detailed local observations, as well as a basis for understanding the values of natural resources and actual practices related to them. By contrast, Western science provides a systematic way of evaluating questions arising from these relationships. Questions pertaining to emerald ash borer and traditional black ash basketmaking being addressed through scientific experiments include: (a) Does submersion of infested black ash logs kill emerald ash borer larvae? (b) How long must logs be submerged for this to happen? (c) Are splints made from submerged wood good enough for basketmaking? and (d) Are traditional submersion practices compatible with federal quarantine regulations?

The emerald ash borer is native to Asia and feeds on the phloem of ash (Yu, 1992). First detected in North America near Detroit, Michigan, in 2002 (Poland and McCullough, 2006), it has since killed hundreds of millions of trees and spread to 31 states and 2 provinces (EAB info, 2017). All North American ash species encountered to date are susceptible. Black (*Fraxinus nigra*), green (*Fraxinus pennsylvanica*), and white (*Fraxinus americana*) ash are the most common ash species in the Great Lakes region and are highly vulnerable (Klooster et al., 2014). Adult beetles emerge in late spring and feed on leaves throughout their lifetime. After mating, females deposit eggs in bark cracks and crevices. Larvae feed in the phloem and cambial region, forming serpentine-shaped tunnels under the bark (Bauer et al., 2004) which disrupt translocation, killing trees within two to three years of infestation (Poland and McCullough, 2006). Larval development can take one to two years (Tluczek et al., 2011), after which adults chew out through the bark, leaving characteristic D-shaped exit holes. Adults may fly up to a few kilometers (miles) (Taylor et al., 2010) and are transported long distances through human-assisted movement of infested firewood, nursery stock, logs, and other wood products.

Black ash seed is being collected to help preserve genetic material for future restoration (Benedict and David, 2003; Agricultural Research Service, 2014). Nevertheless, there is great immediate concern about the availability of black ash for basketmaking in the near future to continue and pass on cultural traditions. In addition, there is concern about movement of infested black ash from areas where it is cut to other lands, including tribal lands, where it is pounded and split to make baskets.

Submergence in water is a traditional method of holding black ash logs from cutting until they are needed for basketmaking. Submergence for at least a month was found effective in killing buprestid beetle larvae (Gardiner, 1962). Therefore, submergence of logs may allow preservation of the wood for basketmaking in the near future and may also be an effective treatment for killing emerald ash borer and preventing artificial spread of this devastating invader.

Black ash are cut for basketmaking throughout the year, but some harvesters prefer cutting at certain times. Basket-grade trees often grow in wetlands;

thus access may be easiest in winter when the ground is frozen (Benedict and David, 2003). Other harvesters prefer cutting trees in the spring when buds begin to swell and sapwood moisture content is high, loosening the bark and annual growth rings (Severn, 2010). Trees may be harvested individually and processed promptly as the need arises, or several may be cut at the same time, then buried or submerged to retain moisture and provide material to work with over time (Severn, 2010). Prolonged submergence for longer than a year may cause wood decay (Wetherbee, 1985).

In the Great Lakes region, traditional methods to process black ash for basketmaking involve systematically pounding debarked, whole logs to delaminate the growth rings from one another using a heavy wooden mallet or the back of an axe (Mundell, 2008). An axe blade or knife is used to score a splint, or strip of thin wood, along the true grain. Each splint is pulled from the log, then rolled up. Splints may be coiled and held in airtight containers for later use and further processing before being woven into baskets. Processing techniques and tools vary from tribe to tribe and by individual. Some artisans use a variety of hand and small shop-made mechanical devices to separate growth rings.

Methods

U.S. Forest Service scientists partnered with basket-makers from the Match-E-Be-Nash-She-Wish Band of Pottawatomi Indians of Michigan (the Band) to test whether infested logs could be submerged to kill within-tree life stages of the insect and retain the wood's color and pliability for basketmaking (Poland et al., 2015). The Band's basketmakers, who come from a long line of master basketmakers, ensured that research methods respected traditional practices from tree selection through the study's design and evaluation of submerged logs' suitability. Suitable trees grow in moist soil or marshy conditions with at least 2 to 3 meters (6.6 to 9.8 feet) of straight, clear lower bole that is fairly large in diameter. A small chink is cut out of the trees to assess tree rings, which should be creamy white and approximately the thickness of a U.S. nickel or quarter (~ 2 millimeters [0.08 inches]). Wood must be moist so it is sufficiently pliable and not break when woven. We submerged the logs in running water deep enough to completely cover the logs, minimizing wood rot that may occur in stagnant water. Following submergence and dissection, logs were transported to Band lands, where they were processed in accord with traditional practices.

Three experiments took place at different times of the year and for various periods. For the first experiment, five trees, approximately 25 centimeters (10 inches) in diameter at breast height, infested with overwintering larvae, were felled in southern Michigan in late April 2010. They were cut into logs approximately 65 centimeters (26 inches) long. Eight logs were randomly assigned to five different treatments: (a) unsubmerged control logs, or logs submerged for (b) 1 week; (c) 4 weeks; (d) 10 weeks; or (e) 16 weeks. We submerged the logs on May 10, 2010, in the Red Cedar River in Okemos, Michigan, with a cinder block attached to an eye bolt at one end. A long rope was attached to the other end of each log, which was secured to a tree along the shoreline to anchor it

Figure 11.1a U.S. Forest Service technicians Tom Baweja and Tina Ciaramitaro submerging black ash logs in the river.

Source: Therese M. Poland

and facilitate retrieval. We periodically measured water depth, temperature, and surface flow rate. When logs were retrieved, half were dissected to determine the density of live and dead emerald ash borer individuals of each life stage. The other half were placed in rearing tubes to allow any living emerald ash borer

Figure 11.1b Ed Pigeon pounding logs with back of axe to separate rings of sapwood.
Source: Therese M. Poland

Figure 11.1c Angie Pigeon shaving black ash splints to remove rough fibers.
Source: Therese M. Poland

Figure 11.1d Finished Pigeon family black ash basket.

Source: Therese M. Poland

to complete development and emerge as adults. Insects removed during dissection were held at room temperature for 24 hours and observed for movement to determine if they were alive. After dissection, logs were pounded and peeled into splints using traditional practices, and splints were assessed for color and pliability (Figures 11.1a–d)

For the second experiment, nine infested trees were felled in February 2011 and cut into logs. Six to eight logs were randomly preassigned to one of six treatments: (a) unsubmerged control, or submerged for (b) 10 weeks; (c) 18 weeks; (d) 26 weeks; (e) 52 weeks; and (f) 78 weeks. We submerged logs in early March when water temperatures were near freezing. We dissected control logs and logs submerged for 10 or 18 weeks within 24 hours of removal from the river. Insects removed during dissection were held at room temperature for 48 hours and observed for movement to determine mortality status. The outer bark had begun to decay and fall off logs submerged for 26, 52, or 78 weeks; therefore, we could not dissect them to assess larval survival. All the logs for each treatment duration were pounded and peeled into splints to assess the impact of prolonged submergence on splint properties.

For the third experiment, infested logs were submerged on May 12, 2011, when water temperature was 5.5°C (42°F). Logs were randomly assigned to one of nine treatments: (a) unsubmerged control logs, or logs submerged for (b) 4 weeks; (c) 5 weeks; (d) 6 weeks; (e) 7 weeks; (f) 8 weeks; (g) 9 weeks; (h) 13 weeks; or (i) 14 weeks. Initially, we assigned eight logs per treatment with four logs designated for dissection and four logs for rearing each week from 4 weeks to 9 weeks. However, by week 6, when we still found little mortality of larvae during dissection, we decided to prolong the experiment by pulling only four logs for dissection each week and foregoing rearing, thus allowing those logs to remain submerged for up to 14 weeks. At 14 weeks, we placed the final set of four logs in rearing tubes to determine adult survival and emergence.

For each experiment, we first confirmed that differences in initial attack densities were random, with no significant differences among treatments that could confound or bias the results. We compared the percentage of dead emerald ash borers of all life stages among treatments by a general linear mixed model (PROC GLIMMIX) because assumptions of analysis of variance were violated (PROC UNIVARIATE). Duration of submergence was tested as a fixed effect, and degrees of freedom were determined using the ddfm KR method. The response distribution was beta with the link function set as logit. Differences among treatments were tested with the Tukey-Kramer means comparison procedure. Data for density of adults emerged from logs placed in rearing tubes were transformed by $\ln(y + 1)$ and then tested for normality and homoscedasticity (PROC UNIVARIATE). We compared transformed data among treatments by analysis of variance (PROC GLM) followed by the Tukey HSD multiple comparison procedure. All analyses were conducted using the SAS 9.4 for Windows statistical package (SAS Institute, 2012) with an α-level of 0.05.

Findings

In experiment 1, the density of adults that emerged from logs decreased significantly as the duration of submergence increased ($F = 5.20$; df = 4, 21; $P = 0.005$). Some emerged from logs submerged for 1 or 4 weeks, but none from logs submerged for 10 or 16 weeks (Figure 11.2). On the other hand, the percentage of dead larvae, prepupae and pupae increased from 7 percent in control logs, to 35 percent after 1 week and 100 percent after 10 or 16 weeks of submergence ($F = 10.8$; df = 4, 17; $P = 0.0002$). It appeared that some larvae actually continued to develop during the first four weeks of submergence because no live or dead pupae were found during dissection of unsubmerged control logs or logs submerged for one week. However, we found live pupae in logs submerged for 4 weeks and dead pupae in logs submerged for 10 or 16 weeks (Poland et al., 2015). The color quality of splints was similar for all treatments; pliability of splints did not decrease after 16 weeks of submergence, and we observed no cracking, splitting, or decay of splints. Water temperature was 13.5°C (56°F) at initial submergence and increased to 21.2°C (70°F) within 12 weeks.

For experiment 2, the percentage of dead larvae and prepupae increased from 5 percent in unsubmerged logs to 100 percent in logs submerged for 18 weeks ($F = 21.8$; df = 2, 16; $P < 0.0001$) (Figure 11.3). Water temperature was freezing

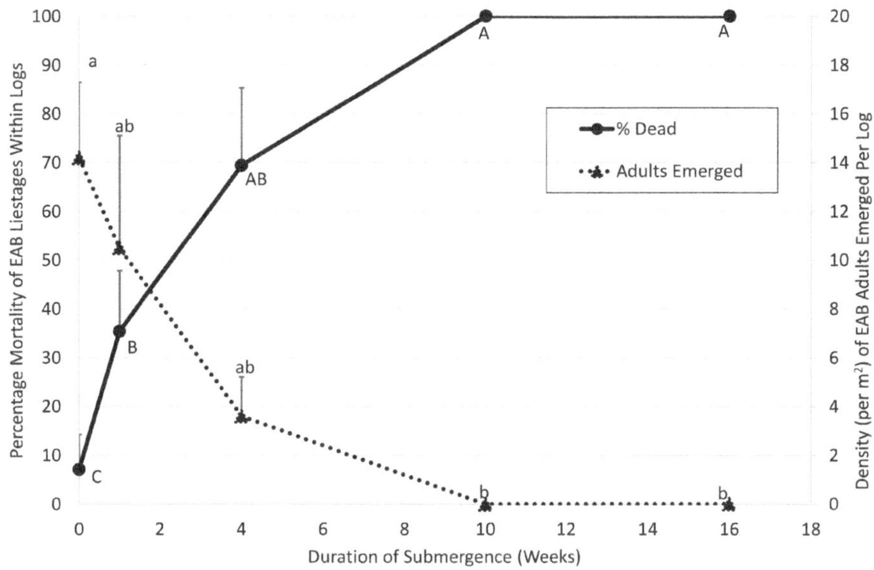

Figure 11.2 Mean (+SE) percentage mortality of emerald ash borer life stages within ash logs and density of adults emerged from ash logs after submergence in the Red Cedar River for various lengths of time in spring/summer 2010. Mean percentage mortality with different upper case letters and mean density of adults emerged with different lowercase letters are significantly different, Tukey multiple comparison procedure, $P < 0.05$. (N = 8 logs per submergence time treatment.)

and remained near 0°C (32°F) for the first 4 weeks, then increased to 5.5°C (42°F) by 2 months, 14.4°C (58°F) by 3 months, 15.5°C (60°F) by 4 months, 21.1°C (70°F) by 5 months, and 25.5°C (78°F) by 6 months, and then gradually decreased back to freezing by 12 months. The bark and first few outer rings of sapwood had begun to decay on logs submerged for 26, 52, or 78 weeks. Most bark had fallen off the logs, and large pieces of decaying bark fell off when they were removed from the river. Therefore, we could not dissect these logs to recover dead larvae, prepupae, and pupae; undoubtedly, they had all died. The inner rings of sapwood were sound, and we pounded and peeled splints from the logs submerged for 26, 52, and 78 weeks in the same manner as those dissected after shorter submergence. Although the outer one to three rings of sapwood began to decay on logs submerged for 26 weeks or longer, color and pliability of splints from the inner rings of sapwood did not decrease even after 78 weeks of submergence; we observed no cracking, splitting, or decay. Logs submerged for at least 26 weeks had a mild pungent odor that completely dissipated when the bark and outer rings of sapwood were removed and the inner rings were exposed to air for a few hours.

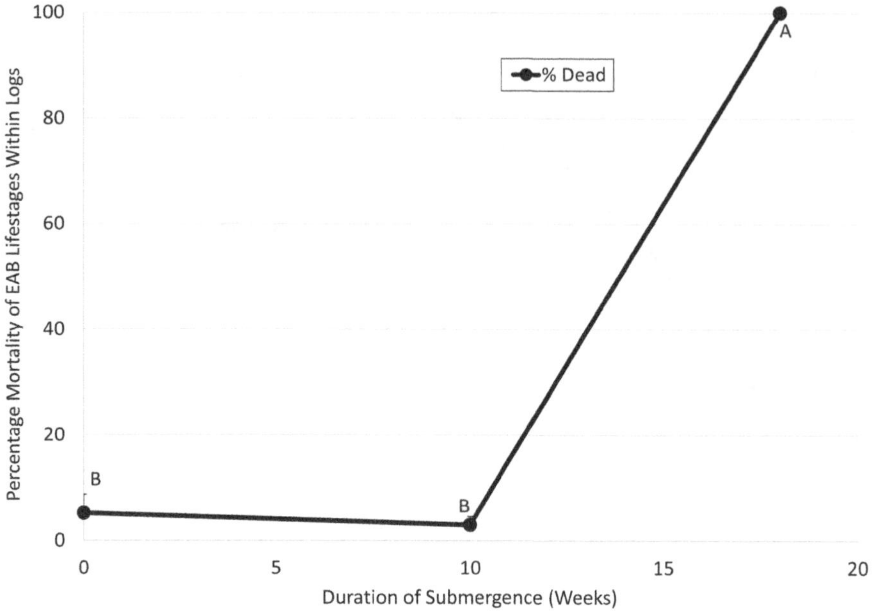

Figure 11.3 Mean (+SE) percentage mortality of emerald ash borer life stages within ash logs after submergence in the Red Cedar River for various lengths of time in winter/spring 2011. Mean percentage mortality with different uppercase letters are significantly different, Tukey multiple comparison procedure, $P < 0.05$. (N = 8 logs per submergence time treatment.)

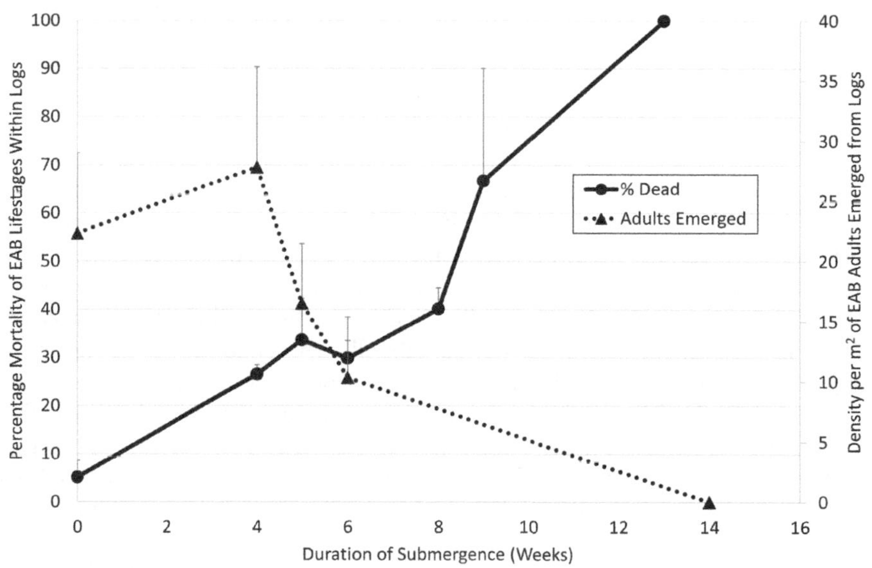

Figure 11.4 Mean (+SE) percentage mortality of emerald ash borer life stages within ash logs and density of adults emerged from ash logs after submergence in the Red Cedar River for various lengths of time in spring/summer 2011. (N = 4 logs per submergence time treatment.)

For experiment 3, the percentage of dead emerald ash borers increased from 5 percent in unsubmerged control logs to 26 percent after 1 week and 100 percent after 13 weeks ($F = 3.25$, df = 7, 23; $P = 0.01$). Some live adults emerged from logs submerged for up to six weeks. No logs were placed in rearing tubes between 7 and 13 weeks after submergence. No adults emerged from logs submerged for 14 weeks (Figure 11.4). Water temperature was 5.5°C (42°F) at submergence and increased to 21.1°C (70°F) by 12 weeks.

Discussion and conclusions

Submerging infested black ash logs effectively killed larvae and prepupae. However, submergence time required for complete mortality varied among experiments and likely depended on time of year, water temperature, and other factors. All larvae, prepupae, and pupae had died, and no adults emerged from logs submerged for 10 weeks during spring 2010 in experiment 1. On the other hand, some larvae survived in logs submerged for 10 weeks during winter 2011 in experiment 2. We did not find complete mortality until 18 weeks of submergence. Similarly, in experiment 3 we did not find complete mortality until 13 weeks of submergence in spring 2011. Temperature conditions may have influenced survival times in submerged logs. Water temperatures were much warmer (starting at 13.5°C [56°F] and warming to 21.2°C [70°F] within 12 weeks) in spring 2010 during experiment 1 compared to winter 2011 and during experiment 2, when temperatures remained near 0°C [32°F] for the first four weeks and gradually warmed to 21.1°C [70°F] over a five-month period. Initial water temperature (5.5°C [42°F]) was also cooler in spring 2011 during experiment 3 compared to experiment 1.

Colder temperatures were found to prolong survival of immersed Douglas fir beetle adults, *Dendroctonus pseudostugae*, (Coleoptera: Curculionidae: Scolytinae), which were able to withstand up to 1,000 hours (5.9 weeks) at 4.5°C (40°F) but only 256 hours (1.5 weeks) at room temperature (Johnson, 1967). When logs are submerged in winter, larvae and prepupae are dormant and acclimated to the cold. They may be able to maintain dormancy and prolong survival while water temperatures remain near freezing. On the other hand, when logs are submerged in spring, larvae have higher metabolic activity and may actively feed and develop. Siegert et al. (2014) found that emerald ash borer larvae feeding in the phloem experienced higher mortality after shorter periods of submersion than those in prepupal chambers.

Insects have a number of adaptations to survive stressful conditions such as hypoxia and immersion (Hoback and Stanley, 2001). Adaptations include altering behaviors, enlarging tracheal system volumes, switching from aerobic to anaerobic metabolic pathways, reducing basal metabolism, and entering a quiescent or dormant state (Hoback et al., 1998; Hoback and Stanley, 2001). Enclosure within a substrate may protect insects from direct contact with water, thereby prolonging survival. Siegert et al. (2014) found that moisture content of the outer 1.3 centimeters (0.5 inches) of ash sapwood increased with duration of submergence. Sapwood moisture doubled from 42 percent before submergence to 81 percent at 12 weeks and 83 percent at 24 weeks after submergence.

The structure of phloem, composed of fibers and sieve tubes with a rapid sealing mechanism in response to stress or wounding (Cronshaw, 1981), may provide some protection from total saturation.

The color quality and pliability of splints did not change significantly with duration of submergence and remained within desirable standards for basketmakers. It is traditional practice to shave rough fibers from the surface of splints to create a brighter and smoother surface. Even after 78 weeks of submergence, splints from all but the first to third outermost rings of sapwood were bright and light-colored. Splints from inner rings also remained pliable after prolonged submergence, without visible splits, crumbling, or decay; however, the outer few rings began to decay after 26 weeks (Poland et al., 2015).

The results suggest that submergence of infested logs during winter or spring and holding them underwater for 13 weeks after water temperatures have warmed above 10–13°C (50–55°F) should ensure complete mortality of emerald ash borer larvae and prevent emergence of adults. These submergence times and conditions preserve wood properties for basketmaking and are compatible with traditional cultural practices.

This study demonstrates how integrating traditional ecological knowledge with scientific ecological knowledge contributes to more sustainable, productive, and locally accepted natural resource management. The emerald ash borer not only threatens forest ecosystems, but also has profound potential implications for American Indian and First Nations communities. The practice of co-managing natural resources has increased in recent decades due to recognition that interagency cooperation and local stakeholder empowerment improve ecological and social outcomes (Conley and Moote, 2003; Bussey et al., 2016). The traditional practice of submerging black ash logs can achieve the immediate goals of providing wood for basketmaking while also killing the emerald ash borer before it spreads. The results may provide guidelines to reduce its inadvertent spread and reduce the threat to black ash, Native American cultural traditions, and livelihoods of basketmakers.

Acknowledgements

We thank Monte Davis, the former environmental specialist of the Match-E-Be-Nash-She-Wish Band of Pottawatomi Indians of Michigan (Gun Lake), for guidance on traditional practices and assistance in planning the study. We thank Tom Baweja and Kailash Brodeur for assistance with field experiments, John Stanovick for assistance with statistical analyses, Damon Crook for evaluation of splint color quality, and Robert Haack and Melody Keena for comments on an earlier version of the manuscript.

References

Bauer, L. S., Liu, H., Haack, R. A., Gao, R., Miller, D. L., and Petrice, T. R. (2004). Emerald ash borer life cycle. In: Mastro, V., and Reardon, R. (Compilers), *Proceedings of the Emerald Ash Borer Research and Technology Development Meeting*. Port Huron, MI, 30

September–1 October 2003. Morgantown, WV: U.S. Department of Agriculture, Forest Service Publication FHTET-2004–02, 8.

Benedict, L., and David, R. (2003). Propagation protocol for black ash (Fraxinus nigra Marsh). *Native Plants*: 24–27.

Berkes, F. (1999). *Sacred Ecology: Traditional Ecological Knowledge and Management Systems*. Philadelphia and London: Taylor and Francis, p. 217.

Bussey, J., Davenport, M. A., Emery, M. R., and Carroll, C. (2016). A lot of it comes from the heart: The nature and integration of ecological knowledge in tribal and nontribal forest management. *Journal of Forestry*, *114*(2): 97–107.

Conley, A., and Moote, M. A. (2003). Evaluating collaborative natural resource management. *Society and Natural Resources*, *16*: 371–386.

Cronshaw, J. (1981). Phloem structure and function. *Annual Review of Plant Physiology*, *32*: 465–484.

Danielsen, K. C. (2002). Maniwiigwaase-Gather birch bark. *Great Lakes Indian Fish Wildlife Commission Mazina'igan, Supplement 1*: 8.

Densmore, F. (1974). *How Indians Use Wild Plants for Food & Crafts (formerly titled Uses of Plants By the Chippewa Indians)*. New York: Dover Publications.

EAB Info. (2017). *Emerald Ash Borer*. Available at www.emeraldashborer.info/index.cfm

Emery, M. R., Wrobel, A., Hansen, M. H., Dockry, M., Moser, W. K., Stark, K. J., and Gilbert, J. H. (2014). Using traditional ecological knowledge as a basis for targeted forest inventories: Paper birch (Betula papyrifera) in the U.S. Great Lakes region. *Journal of Forestry*, *112*(2): 207–214.

Erickson, S. (2006). Iskigamizigan (Sugarbush): A sequel to growing up Ojibwe. *Mazina'igan, Supplement 2006*: 1–2, 4–8.

Fuller, D. (2015). *Maine Balsam Fir: A Tree of Many Uses*. University of Maine, Cooperative Extension Publication, Bulletin # 2541. Available at https://extension.umaine.edu/publications/2541e/

Gardiner, L. M. (1962). Survival of wood borers (Coleoptera: Buprestidae) in water-stored logs. *The Forestry Chronicle*, *38*: 459–462.

Haack, R. A., Bauer, L. S., Gao, T., McCarthy, J. J., Miller, D. L., Petrice, T. R., and Poland, T. M. (2006). Within-tree distribution, seasonal development, and host suitability in China and Chicago. *The Great Lakes Entomologist*, *39*: 169–183.

Hoback, W. W., and Stanley, D. W. (2001). Insects in hypoxia. *Journal of Insect Physiology*, *47*: 533–542.

Hoback, W. W., Stanley, D. W., Higley, L. G., and Barnhart, M. C. (1998). Survival of immersion and anoxia by larval tiger beetles, Cincindela togata. *The American Midland Naturalist*, *140*: 27–33.

Johnson, N. E. (1967). The influence of temperature and moisture on the overwintering mortality of the Douglas fir beetle, Dendroctonus pseudotsugae, in Western Washington (Coleoptera: Scolytidae). *Annals of the Entomological Society of America*, *60*: 199–204.

Klooster, W. S., Herms, D. A., Knight, K. S., Herms, C. P., McCullough, D. G., Smith, A., Gandhi, K. J. K., and Cardina, J. (2014). Ash (Fraxinus spp.) mortality, regeneration, and seed bank dynamics in mixed hardwood forests following invasion by emerald ash borer (Agrilus planipennis). *Biological Invasions*, *16*: 859–873.

Kurtz, C. M., Moser, K. W., Hansen, M. H., Gormanson, D. D., Hatfield, M. A., Sowers, P. A., Dockry, M. J., Emery, M. R., Woodall, C. W., Walters, B. F., Domke, G. M., Gilbert, J., and Wrobel, A. (2015). *Forest Resources Within the Lake States Ceded Territories 1980–2013*. USDA Forest Service, Northern Research Station, Newtown Square, PA. Resource Bulletin NRS-96.

Maier, C. T. (2007). Distribution and hosts of Callidiellum rufipenne (Coleoptera: Cerambycidae), an Asian cedar borer established in the Eastern United States. *Journal of Economic Entomology*, *100*(4): 1291–1297.

Meeker, J. E., Elias, J. E., and Heim, J. A. (1993). *Plants Used by the Great Lakes Ojibwa*. Odanah: Great Lakes Indian Fish and Wildlife Commission.

Morgenstern, K. (2006). Plant profile: Birch (Betula spp.). *Sacred Earth Newsletter*, *V*(2). Available at www.sacredearth.com/ethnobotany/plantprofiles/birch.php

Mundell, K. (2008). *North by Northeast: Wabanaki, Akwesasne Mohawk, and Tuscarora Traditional Arts*. Gardener, ME: Tilbury House Publishers.

Poland, T. M., Ciaramitaro, T. M., Emery, M., Crook, D. J., Pigeon, E., and Pigeon, A. (2015). Submergence of back ash logs to control emerald ash borer and preserve wood for American Indian basketmaking. *Agricultural and Forest Entomology*, *17*: 412–420.

Poland, T. M., and McCullough, D. G. (2006). Emerald ash borer: Invasion of the urban forest and the threat to North America's ash resource. *Journal of Forestry*, *104*: 118–124.

Ragenowich, I. R., and Mitchell, R. G. (2006). *Balsam Woolly Adelgid*. USDA Forest Service, Forest Insect and Disease Leaflet, No. 118. USDA Forest Service, Region 6, R6-NR-FID-PR-04-2006, 1–12.

Ritz, S. E. (2012). *Plant guide for hop Hornbeam (Ostrya virginiana)*. USDA Natural Resources Conservation Service, Appalachian Plant Materials Center. Alderson, WV. Available at https://plants.usda.gov/plantguide/pdf/pg_osvi.pdf

SAS Institute. (2012). *SAS/STAT 9.4 User's Guide*. Cary, NC: SAS Institute.

Severn, J. (2010). Black ash basket making tradition weaves generations together. *Tribal Observer*. Available at www.sagchip.org/tribalobserver/article.asp?article=233

Siegert, N. W., Secord, T., and McCullough, D. G. (2014). Submersion as a tactic to prevent emergence of emerald ash borer Agrilus planipennis from black ash logs. *Agricultural and Forest Entomology*, *16*: 321–325.

Taylor, R. A. J., Bauer, L. S., Poland, T. M., and Windell, K. N. (2010). Flight performance of Agrilus planipennis (Coleoptera: Buprestidae) on a flight mill and in free flight. *Journal of Insect Behavior*, *23*: 128–148.

Tluczek, A. R., McCullough, D. G., and Poland, T. M. (2011). Influence of host stress on emerald ash borer (Coleoptera: Buprestidae) adult density, development, and distribution in Fraxinus pennsylvanica trees. *Environmental Entomology*, *40*: 357–366.

U.S. Department of Agriculture, Agricultural Research Service. (2014). *The NPGS Ash Conservation Project*. Available at www.ars.usda.gov/sp2UserFiles/Place/36251200/Ash_Project/ProceduresPage.html

U.S. Department of Agriculture, Natural Resources Conservation Service. (2014). *Black Ash, Fraxinus nigra Marsh*. USDA NRCS Plant Guide. Available at http://plants.usda.gov/plantguide/pdf/cs-fmi.pdf

Wetherbee, M. (1985). Making a basket from a tree: Splints from lack ash in the Shaker tradition. In: Kelsey, J. (Ed.), *Fine Woodworking: On Bending Wood*. Newtown, CT: Taunton Press, 2–5.

Yu, C. (1992). Agrilus marcopoli Obenberger (Coleoptera: Buprestidae). In: Xiao, G. (Ed.), *Forest Insects of China*, 2nd edition. Beijing, China: Forestry Publishing House, 400–401.

Part IV

Public policy

12 Legislating the Great Lakes

Socially constructing water through congressional discourse

Theresa Castor

Introduction

Physical and social worlds are interconnected, so when people attempt to make sense of the physical world, they assign meaning. However, people may develop different meanings for the same entity or phenomena because of culture and prior experiences. Meaning matters because social action is based on what people assume about a situation or a place. McMullen (2008: 5) observed, "the 'meaning' of water – or how people view the 'value' of water – is critical in designing water policy and effective water-management regimes."

There are different ways that people may interact with the Great Lakes, with different intentions and outcomes. This chapter examines U.S. congressional legislation about the Great Lakes from the 1970s to contemporary times through an interpretivist lens. That examination can help us understand how the Great Lakes have been socially constructed through language in the U.S. political arena. The goal is to identify the meanings of water as constructed through Congress. A focus on congressional discourse is important because Congress, by enacting laws related to the Great Lakes, is a barometer to identify key historical/cultural ways of defining them.

This analysis uses a social constructionist perspective with a focus on communication (Bartesaghi and Castor, 2008). Social constructionism originates from sociology (Berger and Luckmann, 1966), and other disciplines have adopted it, including communication, psychology, education, and anthropology, to examine the subjective meanings that individuals and people develop and assign to objects, events, situations, and other people. Significantly, these meanings become, in practice, "reality," and people therefore act based on those meanings.

A key way to analyze socially constructed meanings is to study words and language. While the Great Lakes have a specific, physical being, what they mean – and subsequently what are appropriate ways to act upon the Great Lakes – rest on the meaning(s) that people have for them.

Different cultures hold various meanings for water (Barnes and Alatout, 2012; Orlove and Caton, 2010). Barnes and Alatout (2012: 483) emphasize that "water as a substance . . . does not merely mediate relations between existing social groups, but constitutes a necessary material for the organization of life in late modernity." McMullen (2008) noted that when we think about Great Lakes policy, we must consider at least two intersecting spheres. First, public policy intersects international and interstate policy because it involves agreements among multiple states and provinces, plus negotiations between the United States and Canada. Second, these policies intersect cultural perspectives, given the two countries' involvement in shaping policy.[1] McMullen noted a tension between defining Great Lakes water as a natural resource versus a tradeable good.

Tarlock (2006) identified five views that have emerged over the past two decades: perpetual gift, world heritage ecosystem, natural resource, commodity, and constrained natural resource. Viewing the Great Lakes as a perpetual gift is a perspective stemming from a Native American view and focuses on the lakes' intrinsic value. The view of world heritage ecosystem focuses on the lakes' value as something that "should be maintained in as close to its natural condition as possible for the benefit of present and future generations" (30). This perspective differs from the first in that the lakes' value is interconnected with their value for humans, although changes are viewed "skeptically" (31). The third perspective of the Great Lakes as a natural resource defines the value of the lakes in terms of benefits for human uses: "Followers of this view seek primarily to develop procedures and standards to decide the conditions under which the use should take place" (33). The fourth perspective as commodity views the lakes "as a natural resource to be exploited for human benefit" and "encourages consumptive use" (33). The fifth perspective of constrained natural resource is "premised on the assumption that consumptive and nonconsumptive uses are legitimate but that the former should be carefully assessed and conditioned to minimize interference with the ecosystem" (35).

Tarlock's (2006) analysis is based on examining water diversion debates and includes perspectives from major stakeholders, states, provinces, Native American tribes, Canadian First Nations, and user and environmental nongovernmental organizations (NGOs). Tarlock's analysis is also binational in comparing Canadian and U.S. views. The analysis in this chapter is broadly inclusive of Great Lakes policy, rather than looking exclusively at water diversions, and focuses on the views of one stakeholder, the U.S. government.

Research questions

The key questions for this project are:

RQ1: Over time, how many bills were proposed related to the Great Lakes?
RQ2: What were their key topics?
RQ3: What were the outcomes of the legislation?
RQ4: Based on the legislative texts and outcomes, what can we infer about congressional meanings for the Great Lakes?

Methods

The primary source of data is the Cable-Satellite Public Affairs Network (C-SPAN) Video Library database, which catalogs congressional legislation and associated discussions. The Video Library contains a record of congressional discussions beginning in 1979. This project uses quantitative and qualitative analyses: basic statistical analysis to present an overview of legislation and a discourse analytic approach to examine language use and changes. Discourse analysis refers to a broad category of research with many subspecialties, but in general focuses on use of language (Schiffrin et al., 2008).

First, this study provides a broad overview of legislation through a keyword search referring to bills with the term "Great Lakes." It analyzes the results based on C-SPAN's subclassification for the legislation and outcomes. A key assumption is that if a bill was approved, then this indicates congressional support, and by extension, approval of an accompanying "meaning" for the Great Lakes. The analysis also examines the subtopics of legislation (e.g., memorials, invasive species, and drought relief).

Findings

Overview of proposed legislation

From the 97th Congress (1981–1982) to the 114th Congress (2016), 748 relevant bills were introduced, an average of 42 per year. Several groupings overlapped in companion bills proposed in both the Senate and House. These are indicated as separate bills although their content and wording were nearly identical. The number of bills passed during that period is much smaller (43, or 5.7 percent). This compares with the 2.8 percent passage rate for all bills enacted into law during that period: 5,723 of 205,937.

Relevant bills, proposed biannually, ranged from a low of 16 (97th Congress; 1981–1982) to a high of 70 (110th Congress; 2007–2008). Four sessions had a much higher number than average: 110th Congress (2007–2008): 70; 101st Congress (1989–1990): 64; 109th Congress (2005–2006): 59; 111th Congress (2009–2011): 59. Banner sessions for enactment of Great Lakes–related legislation were: 101st Congress (1989–1990): 10 enacted; 102nd Congress (1991–1992): 7; 106th Congress (2001–2002): 5; 107th Congress (2003–2004): 4; and 110th Congress (2007–2008): 4.

Key topics: quantitative analysis

The C-SPAN Video Library categorizes bills by topics and keywords and used 53 terms and phrases for indexing Great Lakes bills. Chief among them was the term Great Lakes (703 times). The next four were environmental protection (475); government opinion and politics (475); water resource development (330); and marine and coastal research and fisheries (298). Among the 43 enacted bills the leading topics were Great Lakes (42);[2] environmental protection (36); government opinion and politics (29); public lands and natural resources (26); congressional reporting requirements (26); and authorization (26).

Table 12.1 shows the percentage of Great Lakes bills enacted into law compared to those proposed (column 4). A comparison shows the percentage of approved bills to those proposed on a given topic (column 5).

Table 12.1 Topics of Great Lakes bills: proposed versus approved

Topic	*Proposed*	*Approved*	*Percent approved to proposed*	*Percent of bills approved to proposed on topic*
Great Lakes	703	42	6.0%	2.8% (5,723/205,937)[1]
Environmental protection	475	36	7.6%	4.0% (675/16,621)
Government operations and politics	475	29	6.1%	8.2% (3,016/56,403)
Water resource development	330	22	6.7%	3.2% (552/6,658)
Marine and coastal research and fisheries	298	20	6.7%	7.3% (303/4,136)
Public lands and natural resources	274	26	9.4%	7.5% (1,200/15,994)
Congressional reporting requirements	243	26	10.7%	6.7% (1,255/18,663)
Authorization	187	26	13.9%	7.9% (610/7,674)

1 This cell examines the overall percentage of bills enacted into law compared to those proposed regardless of topic.

Bills related to environmental protection were slightly more likely to pass (7.6 percent compared to 4.0 percent). Those related to water resource development and the Great Lakes were much more likely to pass (6.7 percent compared to 3.2 percent).

Although various bills contained the search term "Great Lakes," the Great Lakes were peripheral in many. The following focuses qualitatively on how the wording of bills framed the Great Lakes as an issue of concern. It is organized chronologically based on Congress number.

Key topics: qualitative analysis

101st Congress

Although bills related to the Great Lakes had been proposed before, the first enacted one in the C-SPAN archive passed in the 101st Congress (1989–1991) when 10 such bills were approved. In four, the Great Lakes were a peripheral issue (e.g., HR 3338 on fish hatcheries in South Carolina). In the others, actions related to the Great Lakes were more central. Two involved general protection of the lakes (HR 1465: Oil Pollution Act of 1990 and HR 4323: Great Lakes Critical Programs Act of 1990). The Oil Pollution Act focused on the Great Lakes as a resource to be protected and recognized them as an internationally shared resource with Canada. The Great Lakes Critical Programs Act amended the Federal Water Pollution Control Act by addressing water quality in the lakes. HR 1668 (the National Oceanic and Atmospheric Administration Ocean and Coastal Programs Authorization Act of 1989) authorized appropriations for the Department of Commerce to carry out Great Lakes research.

The other bills are notable for addressing a specific aspect of the Great Lakes. These topics include fish resources (HR 4299: Great Lakes Fish and Wildlife Restoration Act of 1990) and ship navigation and transportation (S 2740: Water Resources Development Act of 1990 and S 2830: Food, Agriculture, Conservation, and Trade Act). These bills treated the lakes as a resource to be protected and valued for enabling other human activities. For example, the Food, Agriculture, Conservation, and Trade Act addressed the Great Lakes as a means of transportation for shipping. That framing highlighted the lakes' role in commerce while emphasizing their protection for other purposes.

HR 5390, the Aquatic Nuisance Prevention and Control Act of 1990, identified a specific threat – "to prevent and control infestations of the coastal inland waters of the United States by the zebra mussel, and other nonindigenous aquatic nuisance species." Its aim was protection of the Great Lakes, which it recognized as an *ecosystem*. Moreover, it acknowledged the role of human actions in creating conditions that endanger the lake ("the discharge of untreated water in the ballast tanks of vessels and through other means results in unintentional introductions of nonindigenous species." Another key concern about the spread of nonindigenous species was economic: "the potential economic disruption to communities affected by the zebra mussel . . . has been estimated at $5,000,000,000 by the year 2000."

102nd Congress

The 102nd Congress passed seven Great Lakes bills. The Great Lakes were mentioned in two bills as a peripheral topic (HR 1415: Hostage Relief/Compensation and HR 6180: Telecommunications Authorization Act of 1992). Two others addressed the lakes under a broader umbrella that dealt with oceans and seas (HR 2130: National Oceanic and Atmospheric Administration Authorization Act and S 1563: National Sea Grant College Program Authorization Act).

The three other enacted bills are significant for the issues and concerns they identified. The purpose of HR 5013 (Wild Bird Conservation Act) was to promote conservation of exotic wild birds. Embedded in the law was the Great Lakes Fish and Wildlife Tissue Bank Act that recognized the lakes as an ecosystem sustaining humans, animals, and plants – not just bodies of water. HR 5617 (Oceans Act) provided congressional approval of a Governing International Fishery Agreement. This bill is classified as relating to "public lands and natural resources." The Great Lakes are involved in several ways. First, it applied to coastal states, including those on the Great Lakes. Second, a section pertains to inspection of barges entering the lakes. Thus, it provides for protection of the Great Lakes because of their status as internationally shared bodies of water.

HR 6167 (Water Resources Development Act) pertains to work by the U.S. Army Corps of Engineers on projects related to national infrastructure. It is categorized under "water resources development." It amends the Nonindigenous

Aquatic Nuisance Prevention and Control Act of 1990 by adding a paragraph to prevent introduction of aquatic invasive species to the lakes through ballast water on vessels entering through the Hudson River.

103rd Congress

The 103rd Congress enacted only one bill related to the Great Lakes (HR 2150), which authorized funding for the Coast Guard.

104th Congress

The 104th Congress enacted two Great Lakes–related bills: HR 1350 (Maritime Security Act) amended the Merchant Marine Act of 1936 to "revitalize the United States-flag merchant marine. . . ." The Great Lakes are mentioned in the context of marine vessels as transportation, highlighting their role as a tool for human use. HR 4283 (National Invasive Species Act) was classified under "environmental protection" and addressed ballast water management. An amendment to the 1990 Nonindigenous Aquatic Nuisance Prevention and Control Act included specific language about zebra mussels and particular areas of the Great Lakes:

(3) the zebra mussel was unintentionally introduced into the Great Lakes and has infested –

   (A) waters south of the Great Lakes, into a good portion of the Mississippi River drainage,
   (B) waters west of the Great Lakes, into the Arkansas River in Oklahoma,
   (C) waters east of the Great Lakes, into the Hudson River and Lake Champlain.

This bill targets the Great Lakes area and protection of the lakes' overall ecosystem. What is interesting is the legislative specificity in explaining the issue (zebra mussels that were "unintentionally introduced" and have now "infested" several specific waters). This bill prioritized protection of the lakes and identified an environmental threat. It also acknowledged the human role in causing the problem, as well as people's capacity to solve it.

105th Congress

Two bills won approval in the 105th Congress: HR 1481 (Great Lakes Fish and Wildlife Restoration Act) amended the Great Lakes Fish and Wildlife Restoration Act of 1990 to "provide for implementation of recommendations of the United States Fish and Wildlife Service contained in the Great Lakes Fishery Restoration Study Report." Obvious from the title, it emphasizes the management of fish as a resource. It also addresses the Great Lakes *Basin* rather than just the Great Lakes, thus recognizing that this ecological system is not confined to the lakes themselves, but includes interconnected land and river areas.

The other approved bill was S 927 (National Sea Grant College Program Reauthorization Act) that specified that the Great Lakes include Lake Champlain. It was consistently inclusive in addressing the triad of "ocean, coastal, and Great Lakes resources."

106th Congress

The Great Lakes could be considered peripheral in three of the five bills enacted during the 106th Congress. Only two directly related to the lakes. HR 999 (Beach Clean-Up) amended the Federal Water Pollution Control Act by including the Great Lakes as "coastal recreation waters." It emphasized the lakes as a human recreational resource, but because the bill is categorized under environmental protection, it also highlights the need to protect the lakes from pollution.

S 835 (Estuaries and Clean Waters Act of 2000) was classified as pertaining to "water resources development" and used language related to environmental protection: "A bill to encourage the restoration of estuary habitat through more efficient project financing and enhanced coordination of Federal and non-Federal restoration programs, and for other purposes." The Great Lakes are mentioned in a definition of estuary:

> The term "estuary" means a part of a river or stream or other body of water that has an unimpaired connection with the open sea and where the sea water is measurably diluted with fresh water derived from land drainage. The term also includes near coastal waters and wetlands of the Great Lakes that are similar in form and function to estuaries, including the area located in the Great Lakes biogeographic region and designated as a National Estuarine Research Reserve under the Coastal Zone Management Act of 1972.

The Great Lakes are also defined as a "biogeographic region," recognizing that their ecosystem extends beyond the lakes' boundaries.

107th Congress

The lakes were a peripheral topic in three of the four bills enacted during the 107th Congress, mentioning them due to geography or as part of the broader umbrella of the legislation. HR 1070 (Great Lakes and Lake Champlain Act) amended the Federal Water Pollution Control Act, authorizing the Environmental Protection Agency "to carry out projects and conduct research for remediation of sediment contamination in areas of concern in the Great Lakes." It was categorized as relating to environmental protection. It addresses "areas of concern" in the Great Lakes and recognizes the negative impact of sediment contamination.

108th Congress

Three relevant bills were enacted during the 108th Congress. One reauthorized the Harmful Algal Bloom and Hypoxia Amendments Act (S 3014). The other two (HR 2443: Coast Guard and Maritime Transportation Act and HR 289: Ottawa National Wildlife Refuge Complex Expansion and Detroit River International Wildlife Refuge Expansion Act) are intriguing because, at first glance, it seems like the Great Lakes were peripheral issues, but both contain significant specific details and arguments.

HR 289 was categorized as pertaining to public lands and natural resources. It expanded the boundaries of two wildlife refuges. Of note are the first four congressional findings on which the bill is based. First, while the refuge areas relate specifically to Lake Erie, the findings noted the interconnectedness of Lake Erie with the other Great Lakes:

> the western basin of Lake Erie, as part of the Great Lakes ecosystem
>
> (A) is the largest freshwater ecosystem in the world; and
> (B) is vitally important to the economic and environmental future of the United States.

By extension, Lake Erie is valuable given the importance of the Great Lakes. It also emphasizes three Great Lakes features to explain their significance: as a freshwater ecosystem, as an economic resource, and as environmental system. The second finding noted the collaboration between the United States and Canada in water quality and fisheries maintenance, thus recognizing the internationally shared nature of the lakes.

The third finding emphasized the Great Lakes as an ecosystem:

> that increased awareness has been accompanied by a gradual shift toward a holistic ecosystem approach that highlights a growing recognition that shoreline areas, commonly referred to as nearshore terrestrial ecosystems, are an integral part of the western basin and the Great Lakes ecosystem.

Significant is the reflection on *how* the Great Lakes have been viewed ("increased awareness") and a recognition of differing ways or attitudes relating to the lakes ("a holistic ecosystem approach"). This finding emphasizes that the ecosystem is not confined to the lakes but extends to "shoreline areas" and "nearshore terrestrial ecosystems."

The fourth finding noted the quantity of freshwater in the Great Lakes: "The Great Lakes account for more than 90 percent of the surface freshwater in the United States." This finding points to the overall importance of freshwater while highlighting its value as a scarce resource.

HR 2443 made appropriations for the Coast Guard with a section pertaining to the Great Lakes. It allowed the secretary of transportation to "designate a National Maritime Enhancement Institute for the Great Lakes region." The decision-making considerations in this section required the secretary to consider "the unique characteristics of Great Lakes maritime industry and trade" and to study "short sea shipping market opportunities." It also directs the secretary to evaluate "markets for foreign trade between ports on the Great Lakes and draft-limited ports in Europe and Africa," examine "environmental benefits of waterborne transportation," and examine other impacts related to shipping in the Great Lakes. Thus, it stresses the lakes' role in shipping and commerce.

109th Congress

The 109th Congress approved two Great Lakes–related bills: S 1346: Michigan Lighthouse and Maritime Heritage Act and S 2430: Great Lakes Fish and Wildlife Restoration Act of 2006. The Great Lakes appeared in the bills by virtue of location (Michigan) in S1346. S2430 provided "for implementation of recommendations of the United States Fish and Wildlife Service contained in the Great Lakes Fishery Resources Restoration Study."

110th Congress

Four relevant bills became law in the 110th Congress, including the Great Lakes-St. Lawrence River Basin Water Resources Compact, also known as the Compact, one of the most significant laws related to the Great Lakes passed in the late twentieth century.

The America COMPETES Act, also known as the America Creating Opportunities to Meaningfully Promote Excellence in Technology, Education, and Science Act (HR 2272) focused on investment in research and development to improve U.S. competitiveness through innovation. The lakes were identified in the part of the bill that created an Ocean and Atmospheric and Research Development program. It acknowledged the "science" involved with the existence of the Great Lakes. Similarly, the National Sea Grant College Program Amendments Act of 2008 (HR 5618) specified the Great Lakes as among the "sea" resources of the Sea Grant college program, along with the oceans and coasts.

The Great Lakes Legacy Reauthorization Act (HR 6460) amended "the Federal Water Pollution Control Act to provide for the remediation of sediment contamination in areas of concern."

In addition to environmental protection, it recognized water diversion as a specific environmental problem that could harm the lakes. The Compact prohibits diversion of water outside of the Great Lakes Basin. Passage was prompted in part by a Canadian company's 1998 attempt to bottle and ship water from the Great Lakes to Asia. Immediate protest followed the attempt and led to a decade-long process to develop the Compact.[3] It was the only successful Senate joint resolution that dealt with the lakes and required approval by both chambers and the president. As an interstate agreement, it received approval first by the eight Great Lakes states before federal approval.

There are important implications regarding how the Compact legislation socially constructed the meaning of the Great Lakes. For example, there is its definition of the ecosystem: "Basin Ecosystem or Great Lakes-St. Lawrence River Basin Ecosystem means the interacting components of air, land, Water and living organisms, including humankind, within the Basin." It recognized a diversity of components as vital *and* interdependent parts of this system. It contains the following relevant findings, with key words and phrases italicized for emphasis:

a The Waters of the Basin are *precious public natural resources shared and held in trust* by the States;
b The Waters of the Basin are *interconnected* and *part of a single hydrologic system*;
c The Waters of the Basin *can concurrently serve multiple uses*. Such multiple uses include municipal, public, industrial, commercial, agriculture, mining, navigation, energy development and production, recreation, the subsistence, economic and cultural activities of native peoples, Water quality maintenance, and the maintenance of fish and wildlife habitat and a balanced ecosystem. And, *other purposes are encouraged, recognizing that such uses are interdependent and must be balanced*;
d *Future Diversions and Consumptive Uses* of Basin Water resources have the potential to *significantly impact the environment, economy and welfare* of the Great Lakes-St. Lawrence River region;
e *Continued sustainable, accessible and adequate Water supplies* for the people and economy of the Basin are of vital importance; and,
f The Parties have a shared *duty to protect, conserve, restore, improve and manage the renewable but finite Waters of the Basin for the use, benefit and enjoyment of all their citizens, including generations yet to come*. The most effective means of protecting, conserving, restoring, improving and managing the Basin Waters is through the *joint pursuit of unified and cooperative principles, policies and programs mutually-agreed upon*, enacted and adhered to by all Parties.

These congressional findings highlight the lakes as a complex, multifaceted ecosystem that is highly interdependent with human social systems. There are *multiple* uses of their water. The findings recognize the lakes' importance as a freshwater source while recognizing the importance of freshwater (finding 'e'). The Compact also has forward-looking language that considers uses of the waters for future generations and considers the impact of future actions, including diversions and consumptive uses. In terms of how stewardship should be carried out, it emphasizes working together (finding 'f'): "joint pursuit," "unified," "cooperative," "mutually-agreed upon." The Compact also has implications for how decisions about water management should be carried out. For example, one provision focuses on science:

> The Parties commit to provide leadership for the development of a collaborative strategy with other regional partners to strengthen the scientific basis for sound Water management decision making under this Compact.

111th–114th Congress

No relevant bills passed in the 111th Congress. In the 112th Congress, one bill, in which the lakes were peripheral, passed (S404: A bill to modify a land grant patent issued by the secretary of the interior). The 113th Congress passed two relevant bills. S 23 (Sleeping Bear Dunes National Lakeshore Conservation and Recreation Act) involved the shore of Lake Michigan. S 1254 (Harmful Algal Bloom and Hypoxia Research and Control Amendments Act) addressed a specific environmental problem – harmful algal blooms and hypoxia – and defined U.S. coastal waters as including the Great Lakes.

Discussion and conclusion

This analysis of congressional legislation examined the social meanings of the Great Lakes and could be replicated for Canadian legislation. It showed how meanings of the Great Lakes are multifaceted. These meanings emphasize the lakes as an ecosystem, as a means of transportation, as something to be protected, as a natural resource, and more. By looking at the meaning of the Great Lakes, rather than focusing on the more narrow idea of "meanings of water," we see a shift in conceptualization of the lakes. Rather than privileging water, congressional policy has shifted to an ecosystem perspective. In addition, "threats" have ranged from generalized to more specific in identifying a variety of human-related perils such as pollution, invasive species, and water diversions.

Notes

1 While First Nations and Native American groups have provided input on Great Lakes policies and decisions, ultimately the policies are established by the two countries, their states and provinces.

2 The C-SPAN Video Library archive keyword search included one bill with no identifiable connection with the Great Lakes. This likely accounts for the differences of 43 in the keyword search compared to 42 bills categorized by the topic "Great Lakes." This 43rd bill likely was miscoded in the C-SPAN database.

3 See Castor (2016) for an analysis of the congressional discussion on the Great Lakes Compact.

References

Barnes, J., and Alatout, S. (2012). Water worlds: Introduction to the special issue of Social Studies of Science. *Social Studies of Science*, *42*(4): 483–488.

Bartesaghi, M., and Castor, T. (2008). Social construction in communication re-constituting the conversation. In: Beck, C. (Ed.), *Communication Yearbook*. New York: Routledge, 1–39.

Berger, P. L., and Luckmann, T. (1966). *The Social Construction of Reality: A Treatise in the Sociology of Knowledge*. New York: Anchor Books.

Castor, T. (2016). Discursively constructing the Great Lakes freshwater. In: Browning, R. (Ed.), *Advances in Research Using the C-SPAN Archives*. West Lafayette, IN: Purdue University, 33–57.

McMullen, G. J. (2008). *A policy tipping point and the meaning of water in the Great Lakes Basin: Evolving Canada-United States Great Lakes governance in an era of climate change*. Unpublished Master's thesis, University of Wisconsin-Green Bay.

Orlove, B., and Caton, S. C. (2010). Water sustainability: Anthropological approaches and prospects. *Annual Review of Anthropology*, 39: 401–415.

Schiffrin, D., Tannen, D., and Hamilton, H. E. (eds.). (2008). *The Handbook of Discourse Analysis*. Malden, MA: John Wiley & Sons.

Tarlock, D. (2006). Five views of the Great Lakes and why they might matter. *Minnesota Journal of International Law*, *15*: 21.

13 Conservation authorities in Ontario

Key players in the governance of invasive and endangered species

Andrea Olive

Introduction

There is no question that both endangered and invasive species are serious concerns in the Great Lakes region. There are upwards of 250 species at risk[1] in the area under the protection of the U.S. government, the Canadian government, and the provinces of Quebec and Ontario. Examples include the bobolink, the snapping turtle, the Karner blue butterfly, and freshwater mussels. Major threats to regional native species include habitat loss and fragmentation, intensive agriculture, invasive species, and climate change (Francis et al., 2012; Vaughn, 2011).

There are estimated to be more than 180 aquatic and at least 157 terrestrial invasive and non-native species in the Great Lakes ecosystem (Muldoon, 2012). Well-known examples include zebra mussel, round goby, and purple loosestrife. The origins of invasive and non-native species date back to European settlement in the eighteenth and nineteenth centuries. However, as human activity has increased in the region, species are introduced more quickly. Climate change, including warmer average temperatures and greater variance in historical precipitation, has shaped the Great Lakes ecosystem and altered species habitat (Bartolai et al., 2015).

Responsibility for at-risk and invasive species management in the region is complex. For water quality issues, Canada and the United States work together through initiatives such as the Great Lakes Commission and the International Joint Commission. There are also initiatives in the region to co-manage invasive species such as the Great Lakes Panel on Aquatic Nuisance Species, the Asian Carp Regional Coordinating Committee, and the Council of Great Lakes Governors. However, ultimate jurisdiction over flora and fauna remains at the national level. This study examines the local governance of invasive and endangered species, especially the involvement of watershed-based conservation authorities (CAs) in Ontario. Here, governance is operationalized to mean processes, such as decision-making, data collection, elections, and reports, across scale – international, national, subnational, and local (Norman et al., 2013). It is worth briefly describing how each country governs endangered and invasive species within its own borders to contextualize how CAs fit into the overall governance scheme.

In the United States, the Endangered Species Act (ESA) of 1973 applies to all lands, public and private, across the nation. While states can maintain their own endangered species lists, and often have relevant legislation in place, the federal ESA is the supreme law of the land. Thus, if state law and the ESA conflict, the ESA trumps state law. It is essentially the duty of the U.S. Fish and Wildlife Service (USFWS) and National Ocean and Atmospheric Administration (NOAA) to be responsible for the country's flora and fauna. However, the federal government works with many other actors in the process of endangered species protection and recovery, including states, landowners, environmental nongovernment organizations (ENGOs), Native American tribes, and industry stakeholders.

In contrast, invasive species are not the responsibility of any single agency in the United States. Instead, President George W. Bush issued Executive Order 13112 in 1999 to direct all federal agencies that handle invasive species not to carry out or fund activities that could cause or promote the spread of invasive species. The order also created the National Invasive Species Council, which is co-chaired by the departments of Interior, Agriculture, and Commerce (see DOI, 2016a). The council creates four-year management plans for invasive species across the country (see DOI, 2016b). As with endangered species, invasive species governance requires a multitude of actors at the local to international level, as well as inside and outside of government.

The Canadian federal system of government is divided with the federal and provincial governments sharing responsibility for species at risk. The federal government has jurisdiction over federal lands (less than 5 percent of the country outside the Northern territories), migratory birds, and aquatic species. The provinces have jurisdiction over provincial and private lands, as well as wildlife. In 2002, the federal government passed the Species at Risk Act (SARA) that applies to federal jurisdiction and is complemented by policies and regulations across the provinces and territories. In Ontario, the government passed the 2007 Endangered Species Act that looks similar to the American ESA (see Olive, 2014). Responsibility for species at risk falls to the Ontario Ministry of Natural Resources and Forestry (OMNRF), but other actors are involved in the governance of species at risk, such as the federal government, aboriginal bands, landowners, industry, and ENGOs.

Invasive species are co-managed in Canada, but as with species at risk, provinces carry the bulk of responsibility given their jurisdiction over public and private lands and wildlife. There is an Invasive Alien Species Strategy for Canada (see ECCC, 2014), but Ontario passed its own legislation, the Invasive Species Act, in 2014. There are 158 identified invasive species in the provinces, 41 of which are aquatic, like the black carp and European flounder (see Government of Ontario, 2014). There are multiple levels and layers of government and ENGOs involved with invasive species management. The Ontario departments of Environment and Climate Change, and Agriculture, for example, are heavily involved, as well as ENGOs like the Ontario Invasive Plant Council and the Ontario Federation of Anglers and Hunters (see, for example, Connelly et al., 2016).

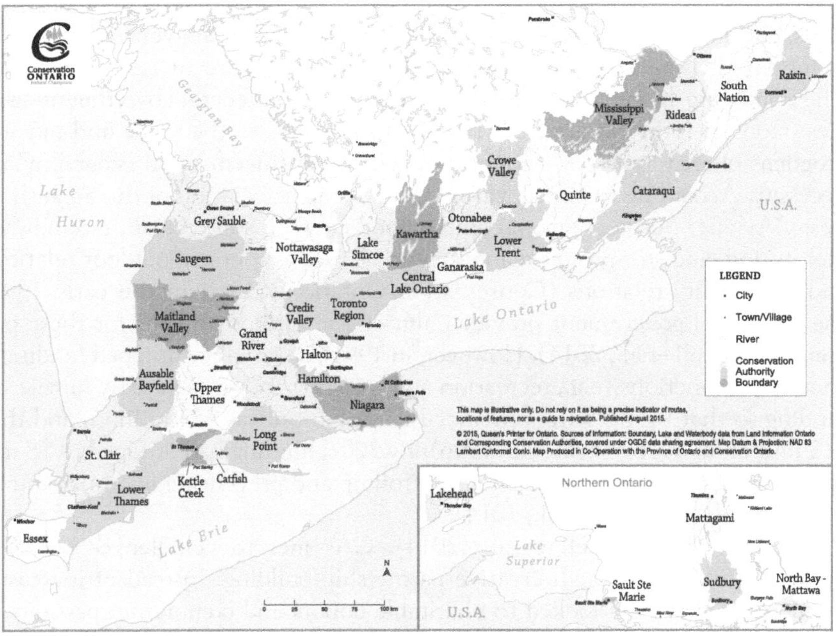

Figure 13.1 Map of conservation authorities in Ontario.

Source: Conservation Ontario and the Queen's Printer for Ontario

This chapter focuses on a unique and understudied actor in Ontario's species governance. Conservation authorities (CAs) were created and are now governed by the Conservation Authorities Act of 1946. Their mandate is to "ensure the conservation, restoration, and responsible management of Ontario's water, land, and natural habitats through programs that balance human, environmental, and economic needs" (Conservation Ontario, 2013). Since water is the main focus of CA responsibility, they operate at the watershed level to protect lakes, rivers, and streams. There are 36 CAs in Ontario – 31 of them in Southern Ontario. See Figure 13.1 for a map. Of the 36 CAs, 26 include a Great Lake shoreline as part of their jurisdiction, and 35 drain into the Great Lakes or St. Lawrence River basins (Conservation Ontario, 2015). While CAs do not map neatly onto Ontario watersheds, there is significant overlap. The Conservation Authorities Act stipulates that a CA can exist only when local residents want to create and fund one. Each CA can exercise jurisdiction over one or more watersheds, but stewardship must involve large-scale planning at that level. From a governance standpoint, CAs "have the power to undertake research, acquire land, raise municipal levies, construct works, control surface water-flows, create regulations, and prescribe fees and permits" (Ivey et al., 2002: 314).

Conservation authorities are run by a board of directors, whose members are appointed by local municipalities. This makes CAs quasi-governmental organizations, but not all their funding comes from government. On average, 48 percent of funding is from municipal taxes while another 40 percent is self-generated. The remaining 12 percent comes from provincial and federal government agencies (Conservation Ontario, 2013b). Funding relates to the scope and range of programs offered by a CA – something that is contingent upon geography and necessity. According to Conservation Ontario, an association of the 36 authorities, CAs offer 30 programs, including flood control, soil erosion, groundwater monitoring, and stormwater management, as well as tourism, outdoor relations, and community relations (Conservation Ontario, 2013c). In the early 1990s, the provincial government provided almost $50 million to CAs for these programs (Mitchell et al., 2014). However, in 1991 the province dropped funding of "non-core" functions (e.g., recreation and tourism) and subsequently further cut funding so that by 1995 the provincial contribution was $33 million, and then $12 million in 1996 (Shrubsole, 1996). By 2002, funding had dropped to $8 million with a priority on projects for controlling and preventing flooding, and for specific lands (Mitchell et al., 2014).

According to Mitchell et al. (2014), CAs met the challenges posed by decreased funding through creative partnership-building. Instead of increasing municipal taxes, CAs looked to charitable donors and community partners, as well as embracing new opportunities like hydropower generation. This means CAs are now able to establish unique and innovative programs while continuing to meet their mandate. However, it also means there is inequity across CAs in terms of the money they can raise and the programs they can subsequently offer. Thus, it is unsurprising that their approach has been labeled "hesitant and unsystematic" (Shrubsole, 1996: 320). CAs have long been criticized for "being unable to deliver consistence resource programmes across the province" (Mitchell et al., 2014: 461). For instance, some offer tourism programs while others do not. Likewise, some may be involved in the removal of invasive species or land acquisition for species at risk, while others may not.

In the past few years, the Ontario government has been reviewing the 1946 Conservation Authority Act, and has signaled interest in modernizing it. In May 2016, the government released a report outlining "proposed priorities for renewal" for CAs. The report's five priorities are accountability, consistency, collaboration, funding, and flexibility (see OMNRF, 2016). While there are major changes afoot for how CAs are funded and what role they play in the province, the report does not discuss program scope or mention species management or conservation. Indeed, there seems to be little acknowledgement of the potential role that CAs could play in species management. Under the 1946 act, CAs can "undertake any scheme for the purposes of conservation, restoration, and development of natural resources other than gas, oil, coal, and minerals." This is a broad mandate – one that may or may not include species at risk or invasive species.

Hypotheses

This chapter asks: What role do CAs play in invasive and endangered species management in Ontario? What connections, if any, and communication do they have with other important actors like government, ENGOs, and landowners? No prior studies have examined their role in the governance of invasive and endangered species. This makes it difficult to test hypotheses stemming from prior research. However, given what is known about expanding role of CAs in environmental policy, this study posits that:

H1: CAs will be very knowledgeable about invasive and endangered species issues in their regions.
H2: CAs will communicate very frequently with the government about invasive and endangered species.
H3: CAs will communicate very frequently with landowners about invasive species and endangered species.
H4: CAs will lack the funding necessary to run invasive and endangered species management programs within their regions.

Methods

In February 2016 we used Survey Monkey to create and send a web survey to all 36 CAs in the province. It contained 35 questions, including yes/no, list, and open-ended questions. All questions were voluntary and a participant could skip any question at any time. The survey instrument contained three main sections: (1) general familiarity with endangered and invasive species, including management and policy, in Ontario; (2) CA involvement with invasive species management, including activities, landowner outreach, and budgetary matters; and, (3) CA involvement with species at risk management, including activities, landowner outreach, and budgetary matters. The overall purpose was to garner new information about the regular activities of CAs and their spending on species management.

The questions required factual or descriptive answers, as opposed to opinion or belief statements. For example, we asked an open-ended question like, "What steps does your Conservation Authority take (if any) to help protect species at risk in Ontario?" No questions involved personal beliefs or attitudes. For example, we did not ask "Should your Conservation Authority take steps to protect species at risk in Ontario?" This distinction is important because we directed the survey to their @info email address and did not specify who within the organization should fill it out. Where there was no @info email address, we sent it to a staff member who specializes in species at risk, invasive species, or landowner outreach. Since there were no attitudinal questions, it is unimportant who filled out the survey at each CA. No demographic information was collected, and the identity of the individual was provided voluntarily only for follow-up questions.

The email to each CA included a letter briefly describing the research, as well as ethics approval and a link to Survey Monkey. After the first email, two reminder emails went to nonrespondents in the following weeks. All responses were coded as yes/no/don't know and entered into a SPSS data spreadsheet.

Findings

Of the 36 CAs in the province, 31 completed the survey. When asked about invasive species in the province, 22 respondents answered that they were "very familiar" with them (see Table 13.1). When asked to name an invasive species in their region, respondents listed a total of 44 species. Phragmites received by far the most mentions (23), followed by garlic mustard (15), emerald ash borer (13), common buckthorn (12), and giant hogweed (10). We cross-checked each named species against the official list of invasive species in Ontario, and no CA mentioned a species that was not listed as invasive in Ontario and/or Canada. Moreover, it is important to acknowledge that CAs were asked only to name a few relevant to their region – and not to provide an exhaustive list. The intent was to examine whether CAs could correctly identify invasive species in their region and to uncover which species are most important to them.

Table 13.1 Respondent familiarity with invasive and endangered species

Question: How familiar are you with invasive/endangered species in your region?	*Invasive Species*	*Endangered Species*
I am not familiar at all	0%	0%
I am very familiar	67%	67%
I am somewhat familiar	28%	30%
I think I might have heard of one or two	3%	3%

Table 13.2 Reported duties and communications for invasive and endangered species

Survey Question	*Yes*	*No*
Is managing issues related to invasive species part of your CAs general duties?	80%	20%
Do you host any meetings or events aimed at educating the public about local invasive species?	69%	31%
Is managing issues related to species at risk part of your CAs general duties?	82%	18%
Do you host any meetings or events aimed at educating the public about local species at risk?	37%	63%

We asked the same set of questions about species at risk. In terms of familiarity, the answers were identical, with 67 percent claiming to be "very familiar" with endangered species. When asked to name species at risk important to their particular region, respondents listed 68. The most frequently named was the Blanding's turtle (14 mentions), followed by the bobolink (11). We cross-checked named species against the official list of endangered species in the province and Canada. In each case, the species was an officially listed endangered or threatened species. Following this line of questioning, we asked CAs if they were familiar with legislation for species at risk. Overall, 75 percent claimed to be "quite familiar" with both the Ontario Endangered Species Act and federal Species at Risk Act.

We asked a series of specific questions about their involvement in invasive and endangered species management (see Table 13.2). Almost all CAs (80 percent) consider managing invasive species to be part of their "general duties." CAs provided a variety of examples, such as monitoring land and water for spread of invasive species and varying degrees of removal, usually determined by budget or manpower restraints. For example, one said it "track[s] the distribution of invasive species in our watershed, and participate in provincial groups" that focus on management. And another admitted that "we have limited resources and work has largely been the removal of plants with partnership groups in the watershed."

Similar for invasive species, 82 percent of CAs said that managing issues related to species at risk is part of their general duties. One said, "The occurrence of species at risk is something we consider in implementing all of our programs." Another provided specific examples: "We run a reptiles at risk program aimed at enhancing or creating habitat to aid and protect species at risk." Another CA provided a laundry list of activities and responded with: "Education. Threat mitigation. Recovery actions based on recovery strategies. Invasive species removal. Work with Conservation Officers. Purchase Property. Work with landowners. Assess properties. Detailed research." Such a list summarizes the work that most CAs claim to do with species at risk in the province.

However, CAs are much more likely to host meetings and events about invasive species (69 percent) than about endangered species (37 percent; see Table 13.2). For invasive species, meetings varied in frequency and were aimed at keeping the public updated on CA yearly goals and teaching the public how to properly control or eliminate invasive species. For example, one CA said it "host[s] a program for high school students to remove garlic mustard and cook with it." Another had "recently hosted an event for municipal staff on creating an invasive species management strategy – seventeen different organizations were represented." For species at risk (SAR), the majority of events were aimed at public schools to try to interest the younger generations about endangered species and habitat conservation. While less than 40 percent of CAs host events, those that do so hold them at least annually. One CA said, "We give workshops and presentations when requested" by other organizations. Another gives "general talks about SAR at tree planting events."

Table 13.3 Reported contact with OMNRF and ENGOs

	Contact with OMNRF	*Contact with ENGOs*
None	0%	3%
Daily	0%	17%
Weekly	57%	43%
Monthly	40%	30%
Yearly	3%	7%

Table 13.4 Frequency of conservation authority communication with landowners

How frequently does a landowner contact you:	*About invasive species*	*About Endangered Species*
Quite frequently	25%	9%
Couple times a month	30%	20%
Couple times a year	42%	46%
Very rarely	3%	25%

We asked how much contact they have with OMNRF and ENGOs regarding species management (see Table 13.3). Overall, there is more frequent contact with ENGOs than with the provincial ministry. That said, all 31 CAs claim to have at least yearly contact with OMNRF, and more than half have weekly contact with the government regarding species management. Almost one-fifth report daily contact with an ENGO in the province, while 43 percent have weekly contact.

Similarly, we asked how often landowners contact CAs regarding invasive and endangered species. Here it appears that landowners are more likely to contact a CA about an invasive species than an endangered species (see Table 13.4). As a follow-up question, we asked why landowners contact a CA about endangered and invasive species. It was common for CAs to claim that landowners contacted them only with concerns about regulations. For example, one said that "usually landowners or real estate agents contact us to know if there are development constraints on their properties related to SAR." Another claimed landowners contact it only because they are "worried what it means if a SAR is found." Similarly, a respondent said landowner contact is "mostly for identification purposes. Also what restrictions will be put on their property should an SAR species be found." For one CA that is rarely contacted by landowners, the respondent said, "It seems that most local landowners are oblivious to SAR and don't usually ask about them. There is the occasional landowner that will ask but it happens rarely."

We asked CAs about their websites as a communication tool with the public (see Table 13.5). Natural habitat protection is featured on all but two CA websites (94 percent total). Invasive species is next most common feature (71 percent), while information about endangered species appears on less than half of

Table 13.5 Conservation authority website information

Does your CA provide information about any of the following on its webpage? (Select all that apply.)	
Natural habitat protection	94%
Invasive species in your region	71%
Wildlife management	43%
Listed endangered species in your region	34%
OESA	26%

websites. Information specific to the Ontario Endangered Species Act is on only one in four websites.

Finally, we asked CAs how much money their agency spends annually on species management. For invasive species, the responses varied greatly. The least amount was $0 while the most was $700,000. Some CAs said their only spending is "staff time," while another said spending "depends largely on external funders." Other CAs said the question was too hard to answer: "Difficult to say. This activity is not a specific program but a portion of general stewardship programming. Perhaps $5,000–7,000 annually." Overall, three said they spend $0, six said they spend between $1 and $5,000, seven spend between $6,000 and $10,000, and the remaining nine spend more than $10,000 annually. Seven did not answer this question.

In terms of spending on species at risk, the answers were similar. The least amount spent was again $0 and the most was $100,000. Again, this proved a difficult question to answer. As one respondent said, it is "hard to put a dollar amount, since many of our projects positively affect all wildlife and humans. Specific SAR research and recovery is a minimum of 70k per year." Another said that "only foundation money is used if we have it," which was echoed by another CA that answered: "none of our own [money] but with grants and support anywhere from 40–60K." Overall, six claimed to spend $0, five spent less than $5,000, five said between $6,000 and $10,000, and eight said they spend more than $10,000 annually. Seven did not answer this question.

Discussion and conclusion

Data collected in this study suggest that Ontario's conservation authorities are working to manage invasive and endangered species in the province. The majority of CAs surveyed consider this management to be part of their general duties. This is surprising since species management is not listed among the 30 programs offered by CAs, and it is not mentioned in the 2016 government review of the mandate of CAs and their work in the province. CAs appear to be significant actors in the governance of Great Lakes invasive and endangered species, but this role is understudied and perhaps unrealized by many government actors and the public at large.

The study hypothesized that CAs are very knowledgeable about invasive and endangered species issues in their regions. Data here suggest that this is largely true. Two-thirds of CAs claim to be familiar with invasive and endangered species. This was supported by their ability to name a wide variety of listed species in their regions. And throughout the survey, respondents provided details about programs they run directed at specific species, which suggests a very high degree of knowledge in these instances.

Second, it was expected that CAs communicate very frequently with the government and ENGOs about invasive and endangered species. More than half have weekly contact with the OMNRF and 40 percent have monthly contact. Respondents also suggested that they often participate in government events and working groups focused on species management. Data-sharing also seems to be part of their relationship with OMNRF, as many CAs mentioned sending reports and information about species to the government. With regard to ENGOS, 17 percent of CAs report daily contact, while 43 percent have weekly contact and 30 percent have at least monthly contact. From the responses it is apparent that CAs work closely with environmental groups to cohost workshops and events across the province. The finding that they communicate frequently with government and ENGOs confirms claims by Mitchell et al. that the "governance model for conservation authorities embodies a blend of 'top-down' and 'bottom-up' approaches to enable dialogue and joint problem-solving at an appropriate scale" (2014: 469). This is a strength of CAs, which respond to local stakeholders but engage in governance with other actors as needed.

The third hypothesis posited that CAs communicate very frequently with landowners about invasive and endangered species. From the data, it appears outreach is occurring, but there is more focus on invasive species than species at risk. This is the case for both the frequency at which landowners contact CAs and the types of workshops and website information provided by CAs. While 82 percent of those surveyed said species at risk management is part of CAs' general duties, only 25 percent of websites provide information on the Ontario Endangered Species Act, and only 34 percent provide information on listed species. In contrast, 71 percent of websites feature information on invasive species, and 69 percent of CAs host events about invasive species.

Finally, the study expected that CAs lack the necessary funding to run species programs within their region. While it is unclear from the responses how much is spent on species management vis-à-vis annual budgets, what is clear is a huge variation in spending. The numbers range from $0 to $700,000 for invasive species and $0 to $100,000 for species at risk. This variation reflects in part by the availability of external funding. This finding is relevant for two main reasons. First, it supports prior research and the 2016 government report that found CA funding/spending to be inconsistent and unreliable. Some CAs have thousands, if not hundreds of thousands, of dollars to spend on species management programs. They can serve local stakeholders well and provide integrated watershed management at an ecosystem level. However, others – for reasons of geography or economic viability in the area – are unable to raise enough external funding to provide local species management programs. Second, this research suggests that if CA involvement in species governance is to continue, then the government (provincial and/or municipal) must provide adequate funding to ensure long-term planning and create some degree of consistency in funding among CAs in the province.

Conservation authorities are not usually included as key players in species governance in Ontario and elsewhere in the Great Lakes region. The 1946 Conservation Authority Act does not specify any role for CAs in invasive species management or endangered species protection and recovery, nor does the 2016 government review of the act. However, CAs have taken up these roles as part of their general duties. And we can assume that their role will only increase as more invasive species enter the region and the number of listed native endangered species continues to rise. Ivey et al. (2002) found that watershed organizations, like CAs, are well-positioned to assess local conditions, track regional trends, identify local issues, foster communication and coordination within governance frameworks, and provide linkages between senior and local levels of government. This study has found that what CAs can do for water management, they may also be able to do with species management.

Note

1 In Canada, the term "species at risk" refers to any species in the categories of endangered, threatened, and special concern. In the United States, the term "endangered species" is commonly used to describe the same categories. For example, the U.S. Endangered Species Act also protects threatened and special concern species. This chapter uses the terms "species at risk" and "endangered species" interchangeably.

References

Bartolai, A. M., He, L., Hurst, A. E., Mortsch, L., Paehlke, R., and Scavia, D. (2015). Climate change as a driver of change in the Great Lakes St. Lawrence River Basin. *Journal of Great Lakes Research*, *41* (Supplement 1): 45–58.

Connelly, A. A., Lauber, R. B., Stedman, R. C., and Knuth, B. A. (2016). The role of anglers in preventing the spread of aquatic invasive species in the Great Lakes region. *Journal of Great Lakes Research*, *42*(3): 703–707.

Conservation Ontario. (2013a). Conservation Authorities of Ontario. Accessed August 17, 2017. Available at http://conservationontario.ca/about-us/conservation-authorities

Conservation Ontario. (2013b). *About Us*. Available at http://conservationontario.ca/about- us/conservation-authorities

Conservation Ontario. (2013c). *History*. Available at http://conservationontario.ca/about- us/conservation-authorities/history

Department of Interior. (2016a). *National Invasive Species Council*. Available at www.doi.gov/invasivespecies

Department of Interior. (2016b). *National Invasive Species Council Management Plan*. Available at www.doi.gov/sites/doi.gov/files/uploads/pre-release_copy_niscmanagement_plan_adopted11july2016.pdf

Environment and Climate Change Canada. (2014). *An Invasive Alien Species Strategy for Canada*. Available at https://ec.gc.ca/eee-ias/default.asp?lang=En&n=1A81B051-1

Francis, C. A, Hansen, T. E., Fox, A. A., Hesje, P. J., Nelson, H. E.,. Lawseth, A. E., and Alexandra English, A. (2012). Farmland conversion to non-agricultural uses in the US and Canada: Current impacts and concerns for the future. *International Journal of Agricultural Sustainability*, *10*(1): 8–24.

Government of Ontario. (2014). *Invasive Species*. Available at www.ontario.ca/page/how-government-combats-invasive-species

Ivey, L., de Loe, R. C., and. Kreutzwiser, R. D. (2002). Groundwater management by watershed agencies: An evaluation of the capacity of Ontario's conservation authorities. *Journal of Environmental Management*, *64*(3): 311–331.

Mitchell, B., Priddle, C., Shrubsole, D., Veale, B., and Walters, D. (2014). Integrated water resource management: Lessons from conservation authorities in Ontario, Canada. *International Journal of Water Resources Development*, *30*(3): 460–474.

Muldoon, P. (2012). Governance in the Great Lakes – A regime in transition. In: Grover, V. I., and Krantzberg, G. (Eds.), *Great Lakes: Lessons in Participatory Governance*. Boca Raton, FL: CRC Press. 44–66.

Norman, E. S., Cohen, A., and Bakker, K. (2013). *Water Without Borders? Canada, the United States, and Shared Waters*. Toronto: University of Toronto Press.

Olive, A. (2014). *Land, Stewardship and Legitimacy: Endangered Species Policy in Canada and the United States*. Toronto, ON: University of Toronto Press.

Ontario Ministry of Natural Resources and Forestry. (2016). *Conserving our Future: Proposed Priorities for Renewal*. Available at apps.mnr.gov.on.ca/public/files/er/conserving-our-future-proposed-priorities-for-renewal.pdf

Shrubsole, D. (1996). Ontario conservation authorities: Principles, practice and challenges 50 years later. *Applied Geography*, *16*(4): 319–335.

U.S. Fish & Wildlife Service Northeast. (2015). Available at www.fws.gov/northeast/nyfo/es/SELIST10-15-2015.pdf

Vaughn, J. (2011). *Environmental Politics: Domestic and Global Dimensions*. Boston, MA: Wadsworth

14 Citizen engagement, interpretation, and resource management in the Great Lakes Basin

Gail A. Vander Stoep

Introduction

The range of natural resources and ecosystems in the Great Lakes biome include lentic and lotic (lake and flowing water) aquatic systems, soils, geological formations, fish and other aquatic species, and terrestrial flora and fauna. All are interrelated components of the Great Lakes Basin impacted by diverse human uses and modifications of ecosystems. Larger-scale events such as climate change are compounding and magnifying more direct human-induced impacts.

Biological and geophysical research is necessary but not sufficient to understand and effectively manage these modified systems. Because humans inherently contribute to management effectiveness – politically, philosophically, economically, and behaviorally – it is crucial to incorporate human dimensions into evidence-based management planning and implementation. This involves engaging multiple stakeholders – residents, industry/business owners, government officials, youth, and community leaders. They hold diverse value systems, levels of awareness, understanding of ecosystems and the underlying biological and physical sciences, and outcome priorities. This mix affects strategies for learning and engagement. Thus, policymakers and resource managers must understand stakeholders to facilitate learning and create opportunities for wide engagement.

Environmental knowledge is fundamental to positive, effective stakeholder engagement in decision-making and stewardship. Knowledge comes from many places: personal experience, formal education, informal learning experiences, and personal and media information inputs. As media channels have diversified and because many people listen primarily to messages consistent with their values and beliefs, it is crucial to understand how and where people access information, what influences their beliefs, and what spurs them to action.

Statement of the problem

The history of resource use, impacts, and management is complex and ecosystem specific, but some generalities have influenced the process of communicating about these resources. Dennis (2001) summarizes major eras of human/nature relationships:

- Nature as provider of resources for basic survival via hunting, gathering, and foraging;
- Wild nature as a "defect" to be tamed and used by humans;
- European ethnocentric views coupled with technology that led to subjugation of Indigenous peoples[1] and colonization of much of the world, partially for extraction of and trade in natural resources;
- In the United States, expansive land clearance and cropping that exploited resources during westward expansion;
- In response to massive losses of resources and environmental disasters, a shift to resource preservation efforts, accompanied by laws and agencies to "preserve" resources; and
- Partial shift to conservation perspectives allowing "wise use" of resources for multiple purposes while protecting them for the long term.

In the early 1990s, resource management philosophies and practices shifted to an ecosystem management approach that regards humans as part of rather than separate from ecosystems. This approach recognizes that ecosystems are dynamic, complex, and require decision-making to consider multiple human and environmental uses, benefits, impacts, and implications across biological, social, cultural, economic, and other systems. An adaptive management approach relying extensively on collaboration and negotiation is more common today (Ecosystem Management Initiative, 2017).

Accompanying these evolving perspectives have been ideas about involving citizens in resource management decision-making. Thus: What factors affect personal and group behaviors that in turn impact resources? What information does the public need and prefer? What are most the effective means of communicating with various publics? And how is information to be used? Shifts in management approaches intertwine with how environmental messages and issues are communicated amidst changing attitudes toward the environment.

Historical grounding of environmental knowledge-sharing

Intimate knowledge of natural species and systems was critical for survival for Indigenous people living close to the land and using resources daily. Thus it was passed from generation to generation by demonstration and oral tradition. During the Middle Ages when European Christianity flourished, only the clergy were allowed to be educated. Because "wilderness" was to be feared, there was little education about or appreciation of natural resources. This attitude influenced early European colonizers in North America. In the 1800s, when resource exploitation was rampant and preservation ideas were emerging, preservation leaders used sketches, paintings, essays, and other writings to promote the grandeur of nature.

By the early 1900s and interwoven with the creation of the U.S. National Park Service in 1916, public programs were primarily entertainment, with tall tales, ghost stories, and other intriguing stories told by lodge owners and stagecoach drivers. Later, based on successful programs by ornithologists at a tourist resort, public programs began to combine biological science with engaging entertainment and encouragement of citizen advocacy for parks. Photographic technology allowed still photos to supplement paintings as visual communicators.

Organized environmental education

After several well-publicized environmental disasters, Rachel Carson (1962) illuminated one example of environmental impacts in her evidence-based book *Silent Spring*. Descriptions of detrimental impacts of DDT and other pesticides spurred debate and shifted attitudes and policy. Public information and environmental protection teaching and advocacy accompanied the first Earth Day in 1970. Concurrently, public schools, government agencies, and nonprofit organizations implemented environmental education programs.

Much public education at the time involved one-way presentation of facts, with a strong dose of advocacy, although school programs often involved students in activities. "Doom and gloom" messages were often the core of communication and education efforts. In response, Hungerford and Lewis (1997) expressed concern that environmental education was "scaring our kids to death."

Partly in response to information-only strategies, nonformal and school-based programs expanded experiential and immersion activities. Guidebooks provided ideas to teachers and nature center directors to get students outdoors to touch, feel, smell, listen, and taste in hopes that positive exposure combined with content knowledge would automatically lead them to "love" and ultimately protect resources.

Early immersion experiences were experiential and fun, but often limited in scientific focus. Many school-based programs were stand-alone courses; curricular add-ons seldom integrated with other subjects. In response, subsequent curricula incorporated experiential learning and linked to government curriculum standards. Nature centers, camps, and other nonprofits developed programs targeting local issues and meeting local curriculum standards. That approach incorporated issues, principles, and human/resource interactions in all subjects, reinforcing the notion that environmental impacts and issues are interrelated.

In the late 1960s and early 1970s, new federal laws such as the Clean Air Act, Clean Water Act, Great Lakes Water Quality Agreement, and National Environmental Policy Act led to significant environmental and health improvements (U.S. Environmental Protection Agency, 2017a, 2017b, 2017c, 2017d). Yet proposals emerge periodically to repeal or weaken these laws (Clean Water Act – Devine, 2015). Although many citizens say they support a healthy environment, they don't take personal, political, and financial actions that support sustainability.

Backlash to environmentalism

So why did the early organized environmental movement fail to win universal acceptance and implementation? And why didn't it result in the intended pro-environmental behaviors? The reasons are numerous. The United States had experienced a backlash to early environmentalism, environmental education approaches, and preservation-based resource management. Doom-and-gloom "environmentalism" messages generated extreme advocacy and prolonged legal battles. Touted catastrophic environmental collapses often did not occur, at least not in easily observable events. Technological and other adaptive measures obscured cumulative impacts that ultimately could lead to environmental tipping points. Instant gratification, convenience, and lower costs often took priority over pro-environmental behaviors and policy support (Vanderheiden, 2011).

Another challenge is an assumption that awareness and knowledge automatically lead to pro-environmental behaviors. However, environmental management and consumer behavior literature illustrate that awareness and knowledge alone do not direct behavior in specific ways – people still eat processed foods and drink soda although they know they are unhealthy; they drive extensively although they know automobiles emit pollutants into the air; they send recyclable products to landfills although they know landfill space is limited and recycling reduces the need for raw materials. Consequently, environmental education philosophies and practices continually shift to try to address challenges.

Changing nature and understanding of resource management issues and goals

Communications challenges are exacerbated because many issues are "wicked problems" – characterized by complexity, a high degree of scientific uncertainty, and a profound lack of agreement on associated values (Balint et al., 2011). Wicked problems – and strategies to address them – often have shades of gray rather than being starkly black-and-white with clear right and wrong choices. Intended and unintended consequences result from management actions. Furthermore, "evidence shows that underlying differences in stakeholder positions are not so much related to uncertainties in science or failure to consider particular aspects of the scientific literature, but rather to conflicting values and preferences" (Balint et al., 2011). Thus, personal and collective management actions are unclear, lacking perfect solutions.

Anticipated outcomes of communication, learning, and resource management also have shifted. Adaptation and resilience are at least as important as strict preservation and protection. Strategies must target people in all walks of life, recognize that human and global systems are dynamic, and consider multiple factors and goals simultaneously. To be effective, communication and management strategies must be relevant to people's values and daily lives.

Communication strategies based solely on scientific information and logic have proven ineffective. More recent strategies acknowledge strong influences of values, different preferred outcomes, and the dynamic nature of contexts. Consequently, we need informed decision-making that considers emotions, values, and multiple world views.

Reflecting this shift, researchers and practitioners have influenced thinking globally. To illustrate, the United Nations Educational, Scientific and Cultural Organization's rationale for its Education for Sustainable Development summarizes changes in communication, education, interpretation, and citizen engagement that also influence changes in resource management:

> Sustainable development cannot be achieved by technological solutions, political regulation, or financial instruments alone. We need to change the way we think and act. This requires quality education and learning for sustainable development at all levels and in all social contexts. Education for Sustainable Development is about enabling us to constructively and creatively address present and future global challenges and create more sustainable and resilient societies.
>
> (UNESCO, 2016)

Method

This chapter reflects a review of literature from social psychology, environmental education, resource management, interpretation, communication, engagement, and cognitive science that demonstrates changes over time in approaches to environmental communication and resources management. Case studies from the Great Lakes Basin illustrate applications of current communication and engagement approaches that facilitate resource management.

Findings

Seeking more effective communication and behavior strategies

Lack of universal support for environmental protection laws and failed strategies have led researchers to explore how to communicate more effective messages about human/environment interactions. The evidence-based realization that exposure and knowledge will *not* automatically lead to pro-environmental behaviors led to new models with diverse variables contributing to learning and behavior. Results identify factors beyond content knowledge that influence learning, engagement, and pro-environmental behaviors. Included are an individual's affective/emotional components, beliefs, and attitudes; influence of significant others; convenience and ease of implementing certain behaviors; specificity of messages; relevancy of outcomes; and personal values. Additionally, brain science has contributed to better understanding of effectiveness of information transmission and processing.

Researchers also have evaluated effectiveness of management approaches, techniques, and underlying behavior models. They have proposed, tested, challenged, modified, and debated several theories and models. Two broad approaches involve:

- *Theory of Reasoned Action* and *Theory of Planned Behavior* (Ajzen and Fishbein, 1970; Ajzen, 1991; de Leeuw et al., 2015; Kaiser et al., 2005), and
- *Norm Activation Model* (Onwezen et al., 2013; Schwartz, 1977).

As both normative (emphasizing norms) and rational cognitive (emphasizing attitudes and beliefs) models have been refined, scholars and practitioners have observed inherent tensions in developing effective interventions to facilitate learning and pro-environmental behaviors. Caveats recognize the complexity of factors influencing behaviors and the changeable nature of internal and external factors that influence different behavioral outcomes. Additionally, intended behaviors must be specific about environmentally sensitive or supportive behaviors.

Developing effective messages for pro-environmental behavior based on models

How do models help determine how best to provide information to facilitate individuals' pro-environmental decisions and actions, particularly if much behavior is influenced by attitudes, values, and perceptions of control? They suggest that information via education and interpretation is important but insufficient. Strategies must understand values and beliefs, interact with individuals based on those values and beliefs, provide skills to act, reduce barriers to perceptions of control over personal behaviors, and engage them directly in learning and action processes. Even more fundamental is to question whether the major purpose of environmental education and interpretation is to shape specific attitudes, beliefs, and behaviors.

Hungerford (2010) summarizes major challenges for environmental education:

- Few educational materials are designed to provide a "rigorous, objective, balanced examination of all sides of environmental issues" (Disinger, 1998);
- Materials designers, promoters, and users are not value-neutral and often actively promote one position or another; and
- Many materials are not grounded in sound science.

Future education materials should address these concerns and, when biased, should clearly present those biases and reasons for them.

Environmental education

By one description, environmental education works at multiple levels to promote sensitivity to environmental issues, provide relevant knowledge, develop attitudes of concern, develop skills to identify and solve problems, train in action strategy knowledge and skills, and participate actively to address challenges (Hines et al., 1987). This is consistent with Hungerford and Volk, who state that the superordinate goal of environmental education is "to aid citizens in becoming environmentally knowledgeable and, above all, skilled and dedicated citizens who are willing to work, individually and collectively, toward achieving and/or maintaining a dynamic equilibrium between quality of life and quality of the environment" (Hungerford and Volk, 1990: 19). Its goal is to facilitate "citizenship behavior."

Informal learning, mindfulness, meaning-making, and interpretation

Falk and Dierking (2000) developed the *Contextual Model of Learning* that highlights the process of meaning-making. Rather than focusing on behavior, it focuses on the meaning that an individual ascribes to an object, event, or experience. It recognizes that learning is cumulative, and that people process new information at the nexus of three contexts: personal context (motivations and expectations, prior knowledge and experiences, prior interests, and degree of choice and control); socio-cultural context (understanding what salient others think, and mediation by the group during the learning process); and physical context (physical and programmatic structure and approach for how new information is presented).

Moscardo (1999) stresses the importance of mindfulness and meaning-making as part of persuasive communication for important messages and supporting pro-environmental behavior. Interpreters can facilitate mindfulness by knowing and respecting the audience, and then creating experiences that actively engage them and subsequently help them better enjoy an experience, process new information, reassess existing perspectives, more effectively make informed decisions, and develop or reinforce a sense of personal control. Interpretive messages should be relevant, present multiple perspectives and outcomes, help people develop their own conclusions, and develop skills to act consistently with those conclusions (Ballantyne and Packer, 2005; Beck and Cable, 2011).

Although brain science and social psychology continue to evolve, results still support traditional principles of interpretation – create relevant messages and experiences that reveal underlying meanings, provoke cognitive and physical engagement, tell comprehensive stories, and connect with the whole of a person (Tilden, 1957; Knudson et al., 2003; Ham, 2012). Those principles can engage visitors in making their own meanings, drawing personal conclusions, and behaving consistently.

Resource management

Just as approaches to interpretation and learning have changed, the philosophy and processes for soliciting and using public input and citizen participation in resource management have changed. Early resource managers applying the perspective of an "expert" model, focused primarily on sharing information, answering questions, and facilitating required public hearings, usually near the end of a planning process. More recently, some managers encourage collaborative planning and decision-making by engaging diverse stakeholders; others suggest empowering citizens to make resource management decisions. Illustrating the range of approaches, the International Association of Public Participation (2014) developed a diagram synthesizing a spectrum of citizen involvement, from passive to active involvement and empowerment. All require sharing of accurate information.

Many believe active citizen engagement ultimately yields the most effective educational and resource management results and reduces litigation. However, these processes are time-intensive and require resource managers to loosen "expert model" perceptions. Furthermore, not all citizens have the time, interest, and resources to commit to such processes. Nevertheless, nonprofit and governmental organizations are shifting to more active citizen engagement through innovative interpretive, participatory, and engagement programs.

From theories and principles to programs: case studies

Evidence-based research can justify techniques and tools for enjoyable and effective learning experiences. However, simple transmission of accurate information, although necessary, is insufficient in people's learning and inclination to engage, or in predicting and influencing specific behaviors. Other relevant factors include values, beliefs, personal characteristics, contextual factors, and beliefs and behaviors of salient others. Interpreters, educators, and managers must engage both the emotional and intellectual parts of the brain and facilitate active processing of messages. Opportunities for socialization and peer support contribute to effective messaging and engagement.

Resource managers can take advantage of effective communication by extending active learning, as these Great Lakes Basin cases illustrate.

Citizen science programs: overview

In the spirit of engaging the public as amateur scientists – to improve science literacy and interest, educate the public about science topics, conduct large-scale and geographically diverse data collection, and facilitate studies that otherwise might not win funding – citizen science moves toward engagement and collaboration among amateurs and professionals. Citizen science occurs at the interface between science and education. Although citizens have collected scientific data for centuries, either from personal passion or to improve their own agricultural productivity, they were marginalized for nearly 150 years as science became professionalized and expert models predominated. Now many scientists and educators see the value of citizen science programs and have created collaborative networks to facilitate such work. Some are large-scale, such as the North American Breeding Bird Survey; others are more locally focused, such as stream monitoring and protection programs (Henderson, 2012; Kight, 2012).

Advances in technology, communications, and transportation facilitate large-scale data collection and centralized data entry. The Citizen Science Alliance (2017), representative of recent growth in such projects, "is a collaboration of scientists, software developers, and educators who collectively develop, manage, and utilise Internet-based citizen science projects in order to further science itself, and the public understanding of both science and of the scientific process. These projects use the time, abilities, and energies of a distributed community of citizen scientists who are our collaborators."

Other projects monitor bumblebees, bird species diversity and migrations, wildlife health, climate patterns and change, butterfly species and migrations, ocean debris, water quality, light pollution, and more. Many operate in the Great Lakes region, several of which are presented below.

Conservation Stewards program

The Michigan Conservation Stewards program (see Figures 14.1 and 14.2) was developed and supported initially by Michigan State University Extension (MSUE). A recent program redesign illustrates changed learning styles and preferences of community members who want to be engaged leaders. It now focuses on community capacity-building and creation of local partnerships. However, to engage in actual conservation work, people need a solid working knowledge of relevant ecosystems and resource management strategies. The original program,

Figure 14.1 Conservation Stewards Program volunteers prepare for a grasslands project.
Source: John Behnke

Figure 14.2 A Conservation Stewards Program volunteer measures a tree.
Source: Winnie Chrzanowski

piloted in 2006, already had moved beyond an older information-intensive model with content delivered primarily by "experts." Although it originally used an "engaged partnership" approach, with professionals and community members as engaged co-learners, it still relied heavily on active, in-person professional leadership and engagement.

The redesign offers more local control and leadership; it shifted from the engaged partnership model to one focused on local community capacity-building. The educational delivery style moved from "binder and PowerPoint content" to blended learning with short videos and online learning exercises. Self-directed learning modules provide the content foundation for field-based activities, use of local presenters, and visits to local sites and organizations during in-class sessions. MSUE shifted its role from lead coordinator to facilitator for local groups to plan and host their own programs. The new model takes an assets-based approach in which local groups identify and use local assets to deliver programs. Although such a model presents challenges – for example, learners' capacity and limited experience with technology – it effectively builds capacity, respects local people as resources, and empowers citizens to remain involved in community-based conservation work (Bhakta and Dann, 2016).

Citizen science programs at Kalamazoo Nature Center, Michigan

The Kalamazoo Nature Center, among the leading U.S. nature centers, is at the forefront of using innovative strategies to extend its mission and reach, and effectively engage diverse publics to protect environmental resources. It offers a variety of citizen science opportunities, and engages people of all ages and levels of science literacy. Currently, it offers opportunities to:

- Tag and monitor monarch butterflies with the Michigan Butterfly Network, which provides teaching materials, instructions, and supplies to create monarch way stations along their migration path, and other conservation engagement opportunities;
- Collaborate locally with the Audubon Society in seasonal bird surveys, winter bird feeder surveys, and Christmas bird counts that contribute to large-scale avian monitoring;
- Sample for aquatic macroinvertebrates in the local watershed in partnership with the Michigan Clean Water Corps that monitors stream health throughout the state;
- Monitor local nest boxes with the large-scale monitoring program coordinated by Cornell University's Lab of Ornithology; and
- Monitor via permanent photo points the health and effectiveness of resource management practices across ecosystems, including prairies, woodlands, and fens (Kalamazoo Nature Center, 2017).

Climate change programs

Climate change programs – both educational and action-generating – are increasing throughout the Great Lakes region. Their goals include communicating scientific evidence and linking it with a range of effects on people's personal and community lives, and providing opportunities for action to mitigate change impacts.

One example is the Michigan Climate Action Network that includes Indigenous peoples working toward stabilizing climate. It provides information and opportunities for individuals, businesses, and communities to: develop sustainable, accessible, and healthy foods; protect ecosystem health with greenways and blueways; increase efficient energy use; create green spaces for native plants; and use stormwater as part of broader water conservation efforts (Michigan Climate Action Network, 2017).

Another technology-linked program is the Great Lakes Climate Change Science and Education Systemic Network, funded by the National Science Foundation. It links education, learning science, and climate science communities. Using data from numerous sources and across diverse ecosystems, it is developing an integrated strategic plan to address climate change. One goal is to engage underrepresented people who are significantly impacted by recent economic downturns and limited educational opportunities (College of Exploration, 2017).

Great Lakes states also have climate change networks, some extensively engaging citizens, others concentrating more on science and education. Still others focus on climate policy development and implementation, while the rest work on resilience and adaptations to climate change. Recognizing the broad and critical nature of climate change education and concentrating heavily on its interpretation and education mission, the National Park Service includes "interpreting climate change," including the ability to use facilitative dialogue and other engagement strategies, as a competency for specialist positions in interpretation (National Park Service, 2017).

5 Gyres in the Great Lakes Basin

Although the nonprofit 5 Gyres Institute is better known for work associated with the plastic pollution crisis in the oceans, it also supports active volunteer partners in the Great Lakes region to address the problem of plastic microbeads that increasingly find their way into the lakes. Its mission is to "empower action against the global health crisis of plastic pollution through science, art, education, and adventure" (5 Gyres, 2017). In addition to expedition and research work, policy work, and partnerships with "cause partners," it created a cadre of support people through volunteers, ambassadors, project partners, and backers who help with local education and action. It provides educational content and demonstration kits to help "ambassadors" tell the stories of plastic's environmental impacts.

Prairie restorations and invasive species control

Small prairie restorations are in progress throughout the region, often by nature centers, land management agencies, other local organizations, and occasionally individuals. These organizations often involve area residents in seed collection and monitoring programs, as well as removal of invasive species – associated directly with prairie restoration or restoring other native ecosystems and habitats, such as savannas. One example is the Stewardship Network that coordinates volunteers organized by numerous geographic "clusters" throughout the Great Lakes region.

Watershed protection and monitoring

Many organizations and initiatives engage local youth in watershed protection, cleaning, and monitoring efforts – all examples of place-based education. The Great Lakes Stewardship Initiative coordinates programs facilitated by nine geographic hubs. Initiated in 2007, its purpose is to develop knowledgeable and active stewards of the Great Lakes through exciting, engaging, hands-on experiences in local communities. One example is the GRAND Learning Network, which "involves mid-Michigan K–12 students and their teachers in school–community partnerships that help them learn about the Grand River watershed region, build their skills and knowledge in science and social studies, and become active stewards of the Great Lakes" (Great Lakes Stewardship Initiative, 2017).

Summary and conclusions

The history of natural resources management, public input and engagement with natural resources management, and educational and interpretive approaches parallels shifts in the purpose and goals of education and resource management. Perspectives about natural resources have evolved from exploitation, through preservation to conservation, and now to adaptive management of ecosystems. Management actions now include human resilience and adaptation.

Early formal resource management relied on professional "experts." Associated and legally mandated public input and information dissemination strategies focused on content delivery by experts. Public meeting input was collected, but ultimate decision-making remained the purview of professionals. Education and interpretation also relied heavily on delivery by experts, eventually with a dose of doom and gloom about environmental issues. In contrast, contemporary trends respect citizens' ability to make informed decisions and take consistent actions. However, we still must facilitate learning opportunities based on the latest science. Effective learning recognizes the importance of knowing audiences, making new content relevant to them and their local contexts, and tapping into each person's emotional and cognitive aspects.

Simply understanding natural systems and impacts and sharing that information with citizens are insufficient for addressing critical resource management issues facing the Great Lakes Basin. Citizens must be involved as co-learners in understanding ecosystems and human roles in and impacts on those systems. They need opportunities for involvement in decision-making. To be effective, they need knowledge, skills, and resource access to make informed decisions about personal behaviors. The process isn't easy. It is time-consuming, involves mutual respect among diverse citizens and professionals, and must recognize the complexity of and interrelationships of environmental issues.

In a highly mediated society, many of us desire quick answers and easy solutions. We often self-select news that supports our values and beliefs. The challenge is developing communications and engagement strategies that address these diverse values and beliefs, emerging challenges and impacts of new technologies, changing environments, and globalization of human impacts on environmental systems. It means recognizing the values and inputs of Indigenous people having relevant perspectives and knowledge to contribute to decisions. It also means actively engaging other underrepresented populations who have been subject to social and environmental injustices.

Note

1 Indigenous peoples includes U.S. federally recognized tribes, First Nations in Canada, and nations and organizations not recognized by the U.S. or Canada as sovereign but that exercise indigenous self-determination.

References

The 5 Gyres Institute. (2017). *The 5 Gyres Institute*. Available at www.5gyres.org

Ajzen, I. (1991). The theory of planned behavior. *Organizational Behavior and Human Decision Processes*, 50: 179–211.

Ajzen, I., and Fishbein, M. (1970). The prediction of behavior from attitudinal and normative variables. *Journal of Experimental Social Psychology*, 6: 466–487.

Balint, P. J., Stewart, R. E., and Desai, A. (2011). *Wicked Environmental Problems: Managing Uncertainty and Conflict*. Washington, DC: Island Press.

Ballantyne, R. R., and Packer, J. M. (2005). Teaching and learning in environmental education: Developing environmental conceptions. *Journal of Environmental Education*, *27*(2): 25–32.

Beck, L., and Cable, T. (2011). *The Gifts of Interpretation* (3rd ed.). Urbana, IL: Sagamore Publishing.

Bhakta, B., and Dann, S. (2016). *Re-designing the Michigan Conservation Stewards Program as a Hybrid Learning Experience: Fostering Conservation Community Capacity-building*. Presentation at the ANREP/NACDEP Joint National Conference, Burlington, VT: June 27.

Carson, R. (1962). *Silent Spring*. Boston, MA: Houghton Mifflin Co.

Citizen Science Alliance. (2017). *What Is the Citizen Science Alliance?* Available at www.citizensciencealliance.org/index.html

College of Exploration. (2017). *Great Lakes Climate Change Science and Education Systemic Network*. Available at www.coexploration.org/greatlakesclimatechange/

de Leeuw, A., Valoi, P., Ajzen, I., and Schmidt, P. (2015). Using the theory of planned behavior to identify key beliefs underlying pro-environmental behavior in high-school students: Implications for educational interventions. *Journal of Environmental Psychology*, *42*: 128–138.

Dennis, S. (2001). *Natural Resources and the Informed Citizen*. Champaign, IL: Sagamore Press.

Devine, J. (2015). New Senate bill would repeal Clean Water Act. *National Resources Defense Council, Expert Blog*. Available at www.nrdc.org/experts/jon-devine/new-senate- bill-would-repeal-clean-water-act

Disinger, J. F. (1998). Tensions in environmental education: Yesterday, today, tomorrow. In: Hungerford, H. R., Bluhm, W. J., Volk, T. L., and Ramsey, J. M. (Eds.), *Essential Readings in Environmental Education*. Champaign, IL: Stipes Publishing, 277–288.

Ecosystem Management Initiative. (2017). *Ecosystem Management Initiative*. Available at www.snre.umich.edu/ecomgt/index.htm and www.snre.umich.edu/ecomgt/emapproach/whatisem.htm

Falk, J. H., and Dierking, L. D. (2000). *Learning from Museums: Visitor Experiences and the Making of Meaning*. Walnut Creek, CA: Altamira Press.

Great Lakes Stewardship Initiative. (2017). GRAND Learning Network. Available at www.glstewardship.org/About/RegionalHubs/TheGRANDLearningNet work.aspx

Ham, S. (2012). *Interpretation: Making a Difference on Purpose*. Golden, CO: North American Press.

Henderson, S. (2012). Citizen science comes of age. *Frontiers in Ecology and the Environment*, *10*: 283. Available at http://onlinelibrary.wiley.com/doi/10.1890/1540-9295- 10.6.283/epdf

Hines, J. M., Hungerford, H., and Tomera, A. N. (1987). Analysis and synthesis of research on responsible environmental behavior: A meta-analysis. *Journal of Environmental Education*, *18*(2): 1–8.

Hungerford, D. (2010). Environmental education (EE) for the 21st century: Where have we been? Where are we now? Where are we headed? *Journal of Environmental Education*, *4*(1): 1–6.

Hungerford, D., and Lewis, B. (1997). Is environmental education scaring our kids to death? *NEA Today*, *16*: 43.
Hungerford, H. R., and Volk, T. (1990). Changing learner behavior through environmental education. *Journal of Environmental Education*, *23*(3): 8–20.
International Association of Public Participation. (2014). *The IAP2 Public Participation Spectrum*. Available at http://c.ymcdn.com/sites/www.iap2.org/resource/resmgr/foundations_course/IAP2_P2_Spectrum_FINAL.pdf
Kaiser, F. G., Hubner, G., and Bogner, F. S. (2005). Contrasting the theory of planned behavior with the value-belief-norm model in explaining conservation behavior. *Journal of Applied Social Psychology*, *35*(10): 2150–2170.
Kalamazoo Nature Center. (2017). *Citizen Science*. Available at https://naturecenter.org/ConservationStewardship/CitizenScience.aspx
Kight, C. (2012). A brief history of Citizen Science. *From Science 2.0*. Available at www.science20.com/anthrophysis/brief_history_citizen_science-93317
Knudson, D., Cable, T. T., and Beck, J. (2003). *Interpretation of Cultural and Natural Resources*. State College, PA: Venture Publishing.
Michigan Climate Action Network. (2017). *Michigan Climate Action Network*. Available at www.miclimateaction.org
Moscardo, G. (1999). *Making Visitors Mindful: Principles for Creating Sustainable Visitor Experiences Through Effective Communication*. Champaign, IL: Sagamore Publishing.
National Park Service. (2017). *Interpreting Climate Change: Competency Description*. Available at http://idp.eppley.org/competencies/specialist/interpreting-climate-change
Onwezen, M. C., Antonides, G., and Bartels, J. (2013). The norm activation model: An exploration of the functions of anticipated pride and guilt in pro-environmental behaviour. *Journal of Economic Psychology*, *39*: 141–153.
Schwartz, S. H. (1977). Normative influences on altruism. In: Berkowitz, L. (Ed.), *Advances in Experimental Social Psychology* (Vol. 10). New York: Academic Press, 221–279.
Tilden, F. (1957). *Interpreting Our Heritage*. Chapel Hill, NC: University of North Carolina Press.
UNESCO. (2016). *Education for Sustainable Development*. Available at http://en.unesco.org/themes/education-sustainable-development
U.S. Environmental Protection Agency. (2017a). *Clean Air Act*. Available at www.epa.gov/clean-air-act-overview/clean-air-act-text
U.S. Environmental Protection Agency. (2017b). *Clean Water Act*. Available at www.thecre.com/fedlaw/legal14water/cwa.htm
U.S. Environmental Protection Agency. (2017c). *Great Lakes Water Quality Agreement*. Available at www.epa.gov/glwqa
U.S. Environmental Protection Agency. (2017d). *National Environmental Policy Act*. Available at www.epa.gov/laws-regulations/summary-national-environmental-policy-act and www.epa.gov/nepa
Vanderheiden, S. (2011). Rethinking environmentalism: Beyond doom and gloom. *Global Environmental Politics*, *11*(1): 108–113.

15 Seven Indigenous principles for successful cooperation in Great Lakes conservation initiatives

Kyle P. Whyte, Nicholas J. Reo, Deborah McGregor, M.A. (Peggy) Smith, James F. Jenkins, and *Kathleen A. Rubio*

Introduction

Academics, governmental and nongovernmental environmental managers, as well as policymakers, often seek practical guidance principles and specific strategies to improve partnerships with Indigenous peoples focused on conservation, ecological restoration, and environmental protection in the Great Lakes region. To develop such principles and strategies, we draw from our study, in which more than 40 representatives from Indigenous-led or -influenced partnerships in the Great Lakes Basin provided diverse ideas about the range of principles that support successful Indigenous participation.

This chapter is based on the largest study to date (Reo et al., 2017) that collectively develops and analyzes case examples of partnerships led or deeply influenced by Indigenous partners. These partnerships focused on outcomes tied to conservation, ecological restoration, and environmental protection in the region. This chapter builds on Indigenous persons' experiences at the ground level of designing partnerships who communicated key principles that made for success or failure in such endeavors. It is an informed exercise in considering the ethics of cooperation in Great Lakes conservation.

We present a practical set of principles for successful partnerships that originated in these projects as expanded through further interpretation based on the author team's collective experiences. We offer informed insights and strategies that those interested in cooperation should consider in their actual or potential partnerships with Indigenous peoples.

Research question

Is there a set of principles to help guide environmental partnerships in the Great Lakes region between Indigenous peoples and national governmental, academic, and nonprofit organizations?

Method

This analysis arises from two research processes. First, during 2012–2014, we interviewed more than 40 participants in multiparty partnerships in the Great Lakes region, which included several hundred Tribes and First Nations, and major Indigenous multiparty organizations such as the Great Lakes Indian Fish and Wildlife Commission and the Chiefs of Ontario. Interviewees represented a cross-case analysis of 39 partnerships involving Tribes and First Nations. The Upper-Midwest Landscape Conservation Cooperative (U.S. Fish and Wildlife Service) funded the project. We looked at partnerships either initiated by Indigenous peoples or in which Indigenous peoples significantly influenced the partnership's organizational structure and practices. Based on an analysis of these examples and focus groups with Indigenous practitioners and their close partners, we developed seven key principles to help cooperative institutions create robust and lasting partnerships. Second, we gathered a group of authors, most with backgrounds in Indigenous conservation, but also from areas as diverse as environmental ethics and history, to interpret the data relevant to guiding principles for collaboration. While protecting the confidentiality of interviewees and discussants, we provide practical distillations of the more philosophically articulated principles.

Findings

Principles are general ideas that help direct planning, policymaking, and interpersonal behavior, subject to adaptation for specific contexts. The principles outlined here are "well-tested" in the sense that the interviewees and authors understood them as having been used effectively – that is, they helped achieve a party's desired goals – in multiple political, cultural, and environmental contexts.

A strategy is an adaptation of a principle for a specific context that seeks to fulfill the spirit of the principle. These participants articulated strategies for enacting the seven principles. Academics, managers, and policymakers should see these principles and strategies as provisional conversation starters for establishing future partnerships and evaluating ongoing partnerships.

Principle of self-determination

This principle concerns Indigenous peoples' right to self-determination. They have this right by virtue of exercising their own governing systems prior to U.S., Canadian, and European colonialism. This right is also enshrined in U.S. laws and policies, such as the government-to-government relationship, and in the United Nations Declaration on the Rights of Indigenous Peoples. Some Indigenous peoples prefer terms such as "sovereignty" or "autonomy" to express self-determination. That preference may be for philosophical reasons or for reasons tied to what linguistic conventions are deemed suitable in certain legal contexts.

Respect for self-determination includes a set of important meanings that vary by social, geographic, and political context.

For most Indigenous peoples, realizing self-determination means that both Indigenous and non-Indigenous governments mutually agree on terms for shared jurisdiction over lands, waters, and resources. Self-determination can mean non-Indigenous parties acknowledge and are bound by formal laws and policies that reference and secure their collective self-governance, as in the case of a treaty or statute. It can also mean ensuring that Indigenous partners are treated as not only citizens, interest groups, or stakeholders of the U.S. or Canada, but also as representatives of self-governing nations in their own right. Indigenous peoples seldom seek full incorporation into the nation-state; at the same time, they rarely seek complete independence or secession (Hendrix, 2008; Napoleon, 2005).

In some cases, self-determination emphasizes cultural difference in that all forms of political or legal order are culturally embedded. That is, self-determination can refer to customary laws and culturally embedded legal orders that guide governance, economic, and social systems (Borrows, 2002). Gender, for example, is important. Indigenous peoples often have different gender systems than settler societies, with implications for how Indigenous leadership is organized.

Given the connections between leadership and gender, many Indigenous peoples interpret discriminatory norms such as sexism as attacks on self-determination and nationhood (Deer, 2015; Goeman and Denetdale, 2009; Whyte, 2016). Indigenous legal orders may ascribe moral concepts to nonhuman entities or beings that do not make sense within the moral frameworks of settlers societies, such as viewing "water" as having personhood, moral responsibilities, or rights (McGregor, 2005). The notion of "respecting self-determination" honors the integrity of Indigenous ways of life that may differ from those of non-Indigenous citizens of a broader, national society. Understandings of self-determination under cultural or customary law ought not be used to sidestep formal treaties or laws protecting Indigenous self-determination. It is nonetheless important to be aware of cases in which the language or terms of treaties or laws was not consensually accepted by the Indigenous parties, hence betraying cultural or customary law in certain respects.

Finally, self-determination sometimes refers to the governmental capacities of specific Indigenous peoples. That is, they should have the bureaucratic/administrative, scientific, legal, and diplomatic capacities to participate equally with state/provincial, federal, and nongovernmental organizations in consultations, jointplanning, negotiations, and other cooperative or adversarial processes. In the United States, for example, Indigenous peoples often press the federal government to ensure their fair or equal access to administrative and programming funds in relation to U.S. states.

Strategies for enacting the principle of self-determination

Non-Indigenous partners should respect self-determination as a governmental, jurisdictional authority. Such respect may extend to regions where Indigenous peoples have historic ties, on lands where they retain treaty-guaranteed resource rights, on lands that were never formally ceded to settler governments (now generally called state and federal governments), or in territories to which they migrated, whether voluntarily or by force. A memorandum of understanding or other formal document should be created to acknowledge their division of authority and responsibility. The advisory, steering, and leadership committees or boards of multiparty institutions should include Indigenous members as representatives of self-determining peoples. In one case, an Indigenous multiparty organization, Anishinabek Nation, created a Women's Water Commission to increase the influence of Indigenous women's leadership roles in water governance in Ontario (McGregor, 2012). Some commission participants are leaders in movements and organizations that exercise self-determination outside of formally recognized Indigenous governments.

Many Indigenous persons saw this representation as an expression of far older forms of governance, stewardship and laws. First Nations have created expressions of their legal orders, such as Chiefs of Ontario's *Water Declaration of the Anishinaabek, Mushkegowuk, and Onkwehonwe* (2008). In partnerships, these expressions propose concepts, protocols, moral codes, traditions, and knowledge that should weigh equally with U.S. or Canadian legal orders. Multiparty institutions should consult with Indigenous groups on issues affecting traditional territory, treaty territory, or title territory. Governmental and nongovernmental organizations owe a duty to inform Indigenous peoples of their plans and actions without Indigenous groups having to inquire to learn about these efforts.

Principle of early involvement

Indigenous peoples should have opportunities to consent or to be consulted early in any process that affects their interests. "Early involvement" or "free, prior, and informed consent" are found in U.S. and Canadian laws and policies and in the UN Declaration on the Rights of Indigenous Peoples. Yet "early" or "prior" can be interpreted many ways. Indigenous representatives should be invited to participate when any partnership is established – "at conception" – to help determine the basic form and operation of the institutional structures. This differs from assumptions that consider it acceptable to reach out only after an institution's or program's specific vision or organizational structure have been outlined by the initial partners, often government agencies or nongovernmental organizations.

Early involvement also respects the legacy that Indigenous peoples historically were not consulted or allowed to say "no" to U.S. or Canadian settlement projects and business interests, including extractive industries and the military. Indigenous peoples' memories, both direct and historical, recall these events as acts of physical violence against them (Deer, 2015; Whyte, 2017a) with wide-ranging ramifications for environmental justice issues such as reproductive justice (Hoover et al., 2012), ceremonial and cultural practices (LaDuke, 1999), and land-use planning that supports flourishing economies and societies (Treaty Indian Tribes in Western Washington, 2011; Walker et al., 2013).

Strategies for enacting the principle of early involvement

Early involvement is best practised through initial meetings where participants discuss their interests in a broadly conceived topic such as "wetlands" or "fishing." After establishing their relevant histories and interests, participants can move into more specific conversations about their capacities to work together. The locations, space, and procedures of initial meetings carry symbolic importance that may affect development of the ensuing institution. It is prudent to recognize this symbolic importance, and Indigenous partners should have the opportunity to host meetings, particularly early in the collaboration. Indigenous peoples often discuss the importance of non-Indigenous parties connecting in person on Indigenous territories to encourage more place-based learning. Early involvement should involve formal expression, such as drafting or documentation of the partnership to ensure that all participants are satisfied with the structure and vision.

These considerations apply to research-oriented partnerships, too. Indigenous partners should participate in planning what kind of research will be performed, how it will be disseminated, and how Indigenous peoples will be involved in interpretation of the data (Smith, 1999).

Principle of intergenerational involvement

Involving all age groups in the cultural dimensions of partnerships is congruent with cultural values in many Indigenous contexts. Indigenous peoples often expect partnerships to bring together different age groups in their community to work meaningfully together and with non-Indigenous collaborators. Partnerships should aspire to involve young people at multiple levels, including but not limited to fieldwork and formal and informal project meetings. Some communities have growing young populations with half or more under 18. Involving youth helps ensure the long-term viability of such efforts. One key consideration involves determining the generations of participants (youth, middle-aged, and elder community members).

The term "youth" is defined broadly to include primary school students through college age and young professionals (generally 25 and under). Prioritizing youth involvement respects the traditional importance of intergenerational relationships. Involvement in these partnerships can help youth learn or reinforce cultural practices and traditional values, can help them develop an appreciation for science, and can help them feel that outsiders respect their cultures. It also motivates, educates, and trains young people to learn about and become good stewards of their environment. Partnerships benefit from the wisdom and insights of a younger generation and from their enthusiasm, optimism, and technical skillsets.

Other segments of the population must also play important roles in many facets of a partnership, especially elders, as well as parents, aunts, uncles, and others. Elders, for example, are respected for their experience-based guidance and wisdom in many Indigenous communities. From the perspectives of some Indigenous peoples, partnerships are an opportunity to involve elders in meaningful work.

Aunts, uncles, and other adults often provide important mentoring and moral support for young people and caretaking for elders.

Strategies for enacting the principle of intergenerational involvement

Younger community members will benefit from engaging in culturally and career-relevant experiences such as event planning, fieldwork, and problem-solving in team environments. Internships provide valuable experiences and substantively contributed to several partnerships. For example, in one case, interns worked alongside a team of scientists and elders who were comparing knowledge on water quality issues affecting an important plant. The youth talked to and interfaced with the elders, building relationships with their community; at the same time, the youth worked with scientists, learning scientific methods for water testing. The youth participated in presenting at conferences both on the water-related findings but also in presentations about weaving together the different knowledge systems. Young people should be encouraged to participate in research and other components of partnerships that have natural linkages to their secondary and postsecondary programs of study.

Youth are also an important target audience for educational and outreach activities created by partnerships. It is important to develop educational materials suited for students at different levels and to have partnership representatives speak to youth audiences inside and outside of schools. Young people can also produce powerful testimonies about the importance of the partnership. Elders should be involved in the decision-making, guidance, and visioning processes of a partnership, whether directly or through an advisory committee.

Many partnerships involved specific roles for linking elders with scientists in performing scientific and technical work, from the design of research to the interpretation of results. In one example, a project to restore an animal population formed a special scientific committee led by several elders who not only played a role in setting the tone and agenda for the meetings but also worked with tribal and nontribal scientists to figure out the best way to do research needed for the restoration work.

Principle of continuous cross-cultural education

Partnership requires that participants have sufficient education and sensitivity about one another's cultural traditions, histories, values, priorities, and aspirations. Yet it may be unclear what the most appropriate mechanisms for cross-cultural learning are or what defines an adequate suite of competencies or level of understanding for engagement with Indigenous peoples. Non-Indigenous partners, including those working for Indigenous governments, are often unfamiliar with the values or do not fully understand the perspectives of Indigenous peoples. For example, many Indigenous peoples are motivated by culturally specific and long-standing – even ancient – responsibilities to care for the land, water, plant and animal populations, and fellow members of one's human community (Bang and Medin, 2010). Cross-cultural differences regarding human–nonhuman relations and perceptions about stewardship are a source of disagreement and misunderstanding in landscape-scale partnerships.

Work done early in a partnership to discuss such cross-cultural differences helps avoid misunderstandings and helps the group overcome unproductive or recalcitrant disagreements (Holtgren, 2013; Ross et al., 2010). Processes of cross-cultural education are not about simply learning "interesting" things about Indigenous cultures. Rather, cross-cultural education is a two-way process in which discussion of cultural traditions opens in-depth conversations about differences in worldview and philosophical areas of overlap.

When working with multiple Indigenous communities, it is important to recognize their cultural distinctiveness. Even among those in the same broader cultural group (for instance, *Anishinaabe*), there are differences in how each community conducts its cultural practices, including those they call upon to conduct ceremonies. Time and human resources should be invested in cross-cultural education, and this process should be an ongoing dimension of the partnership.

Participants never "graduate" from the educational process. One-off sensitivity trainings or primers were seen as insufficient by many people we interviewed and in the authors' experiences. Rarely does brief exposure to guidance actually build strong relationships among partners. Perhaps guidance, such as the principles in this chapter, is best implemented through experience and reflecting on what has been learned through experience.

Strategies for enacting the principle of continuous cross-cultural education

Ongoing cultural liaisons or translators can benefit cross-cultural education. For instance, community members with academic training in Western science or staff with extensive experience bridging Indigenous and non-Indigenous institutions can help avert or resolve conflicts over misunderstandings and cultural differences. But these roles cannot be on top of the work-related responsibilities they already have. That is unfair to those individuals and their communities and nations. Communities themselves should determine who plays such roles. Their traditions, including ceremonies, can be important in partnership meetings and fieldwork and should not be hidden or omitted simply because of a diversity of cultural backgrounds in the group.

The presence of cultural traditions may play an integral part of cooperative group interactions. Ceremonies and other cultural practices lead to cross-cultural conversations, which are perhaps the most important part of this educational process. Non-Indigenous participants learn about Indigenous partners' values and cultural understandings via active participation. Including cultural practices also respectfully informs people outside the partnership that Indigenous peoples are central partners. Indigenous peoples should have space and respect within a partnership to express interest in taking the lead in developing these ideas and practices for cross-cultural learning.

Several partnerships we studied used pipe ceremonies, traditional songs, or convocations by elders to begin meetings or before initiating significant stewardship actions. The Little River Band of Ottawa Indians is well-known for its sturgeon ceremony, a public ceremony designed when the tribe worked with partners to revive sturgeon populations. Both partners and the public come to the afternoon-long ceremony. All attendees participate in cultural activities, such as Indigenous protocols and relationships to sturgeon. At the end of the ceremony, each participant can release a small sturgeon into the river to begin its new life (Holtgren, 2013).

Principle of balance of power and decision-making

Multiparty governance involves many partners who remain formally independent. While this principle may seem obvious, it is important nonetheless, given that Indigenous peoples are sometimes tokenized in partnerships or assumed to lack decision-making authority. Models and structures of decision-making must account for different ways that parties make decisions. Western strategies of decision-making, whether via majority voting or delegation to a committee of experts, may not work for some Indigenous peoples if these strategies are privileged or dominating. Different cultures have different methods for negotiation, conflict resolution, and decision-making (Story and Lickers, 1997).

These methods should be on the table as a partnership is planned. Equal representation alone does not determine the balance of power in a partnership. Individual personalities, decision-making structures, political power, legal rights, and cultural perspectives can all influence the balance of power and group dynamics. Particular personalities or biases (e.g., about the value of different forms of knowledge) can shift the balance of power or dominate discussions. And one or two insensitive or inflexible personalities can ruin a whole initiative.

Strategies for enacting the principle of balance of power and decision-making

Interpersonal relationships must be developed so it is possible to discuss important decisions, including respectful criticism of alternative views. An emphasis on discussion to reach consensus develops relationships among decision-makers and creates understanding. Sometimes ongoing discussions can lead to new relationships, a substantial gain in contexts where Indigenous peoples have not previously worked well with national governments and other partners.

Indigenous peoples may have skills for decision-making that are particularly appropriate in multiparty governance contexts. In one case in Wisconsin, an Indigenous people had a long-standing tradition of "outreach" and "engagement" that their non-Indigenous partners learned from, especially since their approach differed from non-Indigenous approaches that are often assumed to be the only valid ones. In other cases, Indigenous peoples had specific methods of consensus decision-making that were well-suited to their initiative. Haudenosaunee peoples are well-known for using traditional consensus-based decision-making as the organizational basis for making decisions in certain partnerships (Story and Lickers, 1997).

Moreover, Indigenous peoples may make decisions at a different pace than their partners, requiring more time to filter proposals through community channels, for example. Intertribal organizations representing multiple politically distinct tribes often must consult with member governments prior to voting on policies or action items. Partners must recognize that varied paces to make decisions exist, and adjust accordingly. Finally, while Indigenous peoples should be equally represented on management and advisory boards, it may be unrealistic to involve all parties in some contexts. For instance, in the Great Lakes region with well over 100 politically distinct, federally recognized Tribes and First Nations, it is unfeasible to fully involve every Indigenous government in a region-wide initiative. Pragmatism can be balanced with genuine efforts to involve as many relevant groups as possible and to continually acknowledge and communicate with those who are not present.

Principle of respect for Indigenous knowledges

Indigenous societies are guided by their own trusted sources of knowledge. Their knowledge systems are not subsets of scientific knowledge systems; rather, they are knowledge systems engaging in careful and beneficial inquiry and that have developed over many hundreds, if not tens of thousands, of years. Self-determining Indigenous peoples see the practice and revitalization of their knowledge systems as supporting the cohesiveness, cultural integrity, and independence of their communities and nations (Whyte, 2016, 2017b). Indigenous knowledge is valued, then, beyond its role in simply providing useful information.

Multiple Indigenous knowledges can be part of successful multiparty institutions, including elders' knowledge, subsistence practitioners' knowledge, ancestral prophesies, knowledge concerning stewardship responsibilities such as guardianship of lands and waters, knowledge of human–nonhuman relationships, local observations, and ceremonial knowledge. Add to that knowledge of community values, perspectives, priorities, identities, heritages, and worldviews. These knowledges can factor into landscape-scale partnerships in different ways, ranging from understanding cultural protocols and ceremonial practices appropriate for partnership meetings to formulating project priorities to determine research hypotheses. This principle crosses over with the principle of cross-cultural education.

Strategies for enacting the principle of respect for Indigenous knowledges

Indigenous partners must recognize the legitimacy of Indigenous knowledge and respect its potential to influence project priorities and outcomes as equal to their own scientific methods. This is not to say that any source of Indigenous knowledge should be uncritically accepted. Indigenous knowledge can be respectfully scrutinized, and validation is appropriate in some instances, just as scientific concepts and data are scrutinized to ensure validity. In some cases, partners should respect concerns about privacy that may arise when validation requires a community to reveal the location of a sacred place (Williams and Hardison, 2013). Partnerships should begin by respecting *prima facie* Indigenous knowledge and scientific knowledge as equally valid sources of information, insight, and wisdom (Kimmerer, 1998; Whyte, 2013).

The importance of Indigenous knowledges is not confined to their relevance as an additional source of data or evidence in scientific research (Houde, 2007; Latulippe, 2015a, 2015b). While historic and ongoing local observations by Indigenous knowledge-keepers can be valuable, Indigenous knowledges also shape the priorities of a project and bring cultural practices into meetings, research protocols, and decision-making. These knowledge-keepers provide valuable expertise in framing research questions and hypotheses and in providing contextual interpretations of project results. Indigenous partners' communal values will influence their conservation priorities and research ethics (e.g., appropriate ways to handle biological specimens). These are all valid forms of Indigenous knowledge that could be beneficial in the context of partnerships. Given the diversity of Indigenous knowledges, their partners should prepare to interact with diverse knowledge-keepers, as in the case of gender diversity. They should also be open to the possibility that some knowledges filter into the partnership via community members or staff who themselves are not Indigenous knowledge-keepers that is, cultural translators determined by Indigenous communities and nations.

Principle of control of knowledge mobilization

Partnerships addressing landscape-scale environmental issues involve many types and layers of information, knowledge, and research. Knowledge mobilization is the process of bringing together these elements for planning, policy formation, and making decisions. While all partners may have an interest in producing information, sharing knowledge and engaging in cooperative research, they hold different assumptions about how to implement such goals (Smith, 1999; Walker et al., 2013; Whyte, 2016, 2017b). Indigenous peoples are often embroiled in political conflicts over issues such as harvesting rights that require them to disclose reliable information to protect their rights and self-determination. In certain contexts, scientific research can be offensive or threatening to their cultures. Some scientific studies require interactions with species that violate Indigenous ethical standards of conduct for human–animal relations, such as concerns that tracking or monitoring technologies are not ethical ways to interact with certain animals or fish (Williams and Hardison, 2013).

Threats and offense are possible because of differing expectations and concerns about who should generate the data and whether all information resulting from or shared during the project should be publicly available. Community members may feel that their knowledge is being mined for purposes other than their own or that their knowledge will be used against them in development of policies or legislation that does not benefit Indigenous peoples (Dalton, 2010; TallBear, 2013). It is important to think of ways that research can bring value to community initiatives, such as public presentations about community-related issues, developing materials for use in local schools, or shaping the research to provide evidence for projects that Indigenous peoples want to undertake or evaluate.

Strategies for enacting the principle of control of knowledge mobilization

In some instances, it is important for Indigenous peoples to oversee data collection and analysis. Their control legitimizes both the data and the research process in the eyes of the community members they serve. In resource conflicts, Indigenous harvesters often do not trust data generated by states or provinces. In such instances, an Indigenous organization may take on the responsibility of collecting data using both Indigenous and scientific methodologies. In other cases, Indigenous partners may prefer non-Indigenous groups to take charge of data collection. A partnership may wish to give the Indigenous partner resources to develop a research protocol or memorandum of understanding to ensure the community's say in how data are collected and stored, particularly involving traditional knowledge.

These protocols may differ significantly from university institutional review boards (Williams and Hardison, 2013). Some peoples have privacy standards concerning specific types of data or information that are stricter than those of scientists with whom they work. Both Indigenous peoples and institutions such as universities have developed contract language and appendices that express protections of knowledge associated with Indigenous standards of, or concerns about, privacy. Finally, Indigenous organizations can conduct scientific and technical work in ways that support indigenous cultures and self-determination. Although scientific methodologies and technical work may not have been created for the purposes of Indigenous cultures or self-determination, there is no reason why they cannot be redeployed to suit these ends.

Discussion

The principles articulated in this chapter arise from experiences in partnerships led or influenced by Indigenous peoples. While experiences are diverse, many of these principles stem from the value Indigenous peoples ascribe to the ecosystems of their territories in the Great Lakes region. Throughout this project, elders and many others expressed the significance of political, cultural, historical, social, and educational relationships to their territories. While each Indigenous people has a different history of relocation and territorial dispossession, partners should not diminish or downplay the importance of territory for being able to exercise self-determination, practice knowledge systems, and perform ceremonies and social activities. Partners must try to reflect on how to play a part in the work Indigenous peoples do on behalf of Indigenous territories rather than thinking only about how Indigenous peoples can support the goals of non-Indigenous partners.

These principles are not meant to be unwavering rules and are not articulated as such. Rather, they are key heuristics or rough guidelines that should be observed diligently in forming conservation partnerships. While they may be articulated "roughly" here, that does not mean there are no serious consequences and harms associated with failure to observe them. Disrespecting self-determination or violating privacy can set back the potential for present or future collaboration by decades. On the other hand, taking these principles seriously promises a strong possibility of well-founded relationships that move toward higher levels of trustworthiness and mutual respect.

References

Bang, M., and Medin, D. (2010). Cultural processes in science education: Supporting the navigation of multiple epistemologies. *Science Education*, 94(6): 1008–1026.

Borrows, J. (2002). *Recovering Canada: The Resurgence of Indigenous Law*. Toronto, ON: University of Toronto Press.

Chiefs of Ontario. (2008). *Water Declaration of the Anishinaabek, Mushkegowuk and Onkwehonwe*. Toronto, ON: Chiefs of Ontario.

Dalton, Z. K. (2010). *As We Move Ahead Together: Foregrounding Reconciliation and Renewed First Nations/non-Aboriginal Relations in Environmental Management and Research – An Examination of the Species at Risk Conservation and Recovery Scenario in Southwestern Ontario*. Toronto, ON: University of Toronto.

Deer, S. (2015). *The Beginning and End of Rape*. Minneapolis, MN: University of Minnesota Press.

Goeman, M. R., and Denetdale, J. N. (2009). Native feminisms: Legacies, interventions, and Indigenous sovereignties. *Wicazo Sa Review*, 24(2): 9–13.

Hendrix, B. A. (2008). *Ownership, Authority, and Self-Determination*. State College, PA: Penn State Press.

Holtgren, M. (2013). Bringing Us back to the River. In: Auer, N., and Dempsey, D. (Eds.), *The Great Lake Sturgeon*. East Lansing, MI: Michigan State University Press, 133–147.

Hoover, E., Cook, K., Plain, R., Sanchez, K., Waghiyi, V., Miller, P., Dufault, R, Sislin, C., and Carpenter, D. O. (2012). Indigenous peoples of North America: Environmental exposures and reproductive justice. *Environmental Health Perspectives*, 120(2): 1645–1649.

Houde, N. (2007). The six faces of traditional ecological knowledge: Challenges and opportunities for Canadian co-management arrangements. *Ecology and Society*, 12(2): 34–52. [online] http://www.ecologyandsociety.org/vol12/iss2/art34/

Kimmerer, R. (1998). Intellectual diversity: Bringing the Native perspective into natural resources education. *Winds of Change*, 13(3): 14–20.

LaDuke, W. (1999). *All Our Relations: Native Struggles for Land and Life*. Cambridge, MA: South End Press.

Latulippe, N. (2015a). Bridging parallel rows: Epistemic difference and relational accountability in cross-cultural research. *International Indigenous Policy Journal*, 6(2): 1–17.

Latulippe, N. (2015b). Situating the work: A typology of traditional knowledge literature. *AlterNative: An International Journal of Indigenous Peoples*, 11(2): 118–131.

McGregor, D. (2005). Traditional ecological knowledge: An Anishnabe woman's perspective. *Atlantis: Critical Studies in Gender, Culture and Social Justice*, 29(2): 103–109.

McGregor, D. (2012). Traditional knowledge: Considerations for protecting water in Ontario. *International Indigenous Policy Journal*, *3*(3): 1–23.

Napoleon, V. (2005). Aboriginal self-determination: Individual self and collective selves. *Atlantis: A Women's Studies Journal*, *29*(2): 31–46.

Reo, N. J., Whyte, K. P., McGregor, D., Smith, M. P., and Jenkins, J. F. (2017). Factors that support Indigenous involvement in multi-actor environmental stewardship. *AlterNative: An International Journal of Indigenous Peoples*, *13*(2): 58–68.

Ross, A., Sherman, R., Snodgrass, J. G., and Delcore, H. D. (2010). *Indigenous Peoples and the Collaborative Stewardship of Nature: Knowledge Binds and Institutional Conflicts*. Walnut Creek, CA: Left Coast Press.

Smith, L. T. (1999). *Decolonizing Methodologies: Research and Indigenous Peoples*. London: Zed Books.

Story, P., and Lickers, F. (1997). Partnership building for sustainable development: A First Nations perspective from Ontario. *Journal of Sustainable Forestry*, *4*(3–4): 149–162.

TallBear, K. (2013). *Native American DNA: Tribal Belonging and the False Promise of Genetic Science*. Minneapolis, MN: University of Minnesota Press.

Treaty Indian Tribes in Western Washington. (2011). *Treaty Rights at Risk: Ongoing Habitat Loss, the Decline of the Salmon Resource, and Recommendations for Change*. Available at http://nwifc.org/w/wp-content/uploads/downloads/2011/08/whitepaper628finalpdf.pdf

Walker, R., Natcher, D., and Jojola, T. (2013). *Reclaiming Indigenous Planning* (Vol. 70). Montreal, QC, and Kingston, ON: Canada: McGill-Queen's University Press.

Whyte, K. P. (2013). On the role of traditional ecological knowledge as a collaborative concept: A philosophical study. *Ecological Processes*, *2*(1): 1–12.

Whyte, K. P. (2014). Indigenous women, climate change impacts and collective action. *Hypatia: A Journal of Feminist Philosophy*, *29*(2): 599–616.

Whyte, K. P. (2017a). The Dakota access pipeline, environmental injustice and U.S. colonialism. *Red Ink – An International Journal of Indigenous Literature, Arts and Humanities*, *19*(1): 154–169.

Whyte, K. P. (2017b). What do Indigenous knowledges do for Indigenous peoples? In: Nelson, M. K. and Shilling, D. (Eds.), *Traditional Ecological Knowledge: Learning from Indigenous Methods for Environmental Sustainability*. Cambridge: Cambridge University Press.

Williams, T., and Hardison, P. (2013). Culture, law, risk and governance: Contexts of traditional knowledge in climate change adaptation. *Climatic Change*, *120*(3): 531–544.

Part V

Conclusions

16 Through the crystal ball

By Mark Neuzil and *Eric Freedman*

Isle Royale, which sits in Lake Superior and belongs to the United States through the state of Michigan, hosts the most-studied predator–prey animal relationship in the history of the Great Lakes watershed. On the 205-square-mile (530-square-kilometer) island, the largest land mass in a group of more than 400 that make up Isle Royale National Park, scientists have examined moose continuously since at least the late 1920s. The moose's lab partners, wolves, have not been on Isle Royale as long but have been under the microscope since they arrived.

The reasons are straightforward. Isle Royale is a more-or-less contained ecosystem, and its animals have existed in relatively small – nearly countable – numbers, waxing and waning over the years. Although the island sits in the middle of the largest freshwater lake on the continent, it is easily accessible by boat, ski plane, or seaplane unless the weather is terrible. The archipelago is about 29 kilometers (18 miles) from the Minnesota shore, 93 kilometers (58 miles) from Michigan. Scientists also have the advantage of not having to navigate around many tourists; more people visit Yellowstone National Park in a single summer day than trek to Isle Royale in the entire season, which runs from May to September. (Although those who do come stay a while. The National Park Service reports that the average national park visitor stays four hours; on the island, it is closer to four days.) There are no roads and no motors. No one stays there for any length of time over the winter except park employees and scientists.

A less-straightforward, but equally important reason from a communications perspective is that moose and wolves represent what are called charismatic megafauna. That is, they are large, easily recognized, photogenic, popular, furry mammals that end up on the covers of magazines, in wildlife films, newspaper websites, plush toys, and the odd Disney movie. As such, wolves and moose have cultural meanings that go beyond their predator–prey relationship; they are objects of public interest and media attention. The wolf, in particular, gets extra media points for being on and off the U.S. endangered species list. The wolf loses points for those who remember what happened to Little Red Riding Hood, Peter and the Three Little Pigs as objects of the Big Bad One's attention. On the other hand, moose, as prey and as plant-eaters, engender empathy – or at least the cultural favoritism reserved for the underdog. The fact that its population has been in a steep downward curve since the 1990s adds concern.

The researchers who first took extended looks at the moose on Isle Royale included Adolph Murie (1899–1974), who has been called the first scientist to study wolves in their natural habitat (Grooms, 2002). Murie, who was born in Moorhead, Minnesota, teamed with his brother Olaus (1889–1963) to become the most famous sibling biologists in U.S. science history. Their accomplishments are long and varied. That includes pioneering scientific studies, contributing to national park expansions, creating and leading conservation groups, and lobbying and testifying in Congress. A fine writer, Adolph went on to win the 1963 John Burroughs Medal for the year's best nature book.

Adolph's early career for the National Park Service took him to Isle Royale, where in 1934 he produced a 44-page report on the moose. The study to determine "potential future uses of the island" was funded by the Michigan Legislature in 1929. Not much extended scientific work had been done in the island since a 1905 visit by Charles C. Adams, who notably omitted moose in his indexed report of mammals.

Murie was assigned to look at the mammals again. He decided to devote an entire report to the moose and covered all other mammals in a separate document. He landed on the island on July 14, 1929, with his assistant, Paul F. Hickie, and set up camp at Pickerel Cove, starting a legacy of biological research that continues today. The men worked until mid-October, and Murie returned the following May with another biologist, Norman A. Wood, who was assigned to study birds. Hickie was hired by the Michigan Game Division and returned to the area in the winters of 1934–37 with a plan – aided by members of the Civilian Conservation Corps – to trap starving moose and restock the nearby Upper Peninsula of Michigan. Hickie and his team trapped 71 moose, but the scheme failed (Hickie, 1943).

Murie's report, which attracted little attention at the time, included interviews with old-timers who had been watching and/or hunting moose since the 1880s on the island and the mainland. A moose migration may have come in 1912–13 in an unusually cold winter that solidified an ice bridge stretching all the way to Canada. Others probably swam across, perhaps earlier in the century.[1] As wolves had not yet arrived, the moose had "little check on their increase" and had the run of the place (Murie, 1934: 7). Murie guessed more than 1,000 moose lived on Isle Royale in 1929, and thought "that a count would give a figure far greater than the estimated minimum" (Murie, 1934: 10). It would not surprise him, Murie reported, to find 2,000 or 3,000 there. By 1929, the plant life favored by the moose was suffering from over-browsing. He and his assistant shot several moose, including a pregnant female, and measured and weighed them, including the 10-kilogram (22-pound) unborn calf. They also found and examined 24 carcasses and skeletons.

The 1930s proved to be an important decade for the Isle Royale moose. The media found the animals and thrust them into charismatic megafauna status through a photographer named Walter Hastings. An employee of the Michigan Department of Conservation, Hastings' films on the moose of Isle Royale made in the 1920s and 1930s were shown all over the world. Hastings went on to make wildlife films of all sorts, and it is a mark of his legacy that the Audubon Society chapter in Traverse City, Michigan, is named after him.

Worried about overpopulation after considering and rejecting hunting, Murie became the first of many scientists to suggest introducing a predator to the island ecosystem. Shipping live moose – at 363 kilograms (800 pounds) or more each – did not seem a practical option to Murie either, an assessment that was proved accurate a few years later based on Hickie's failed attempts. Instead, Murie suggested bears, cougars, or wolves, since "such an introduction of a native species would add materially to the animal interests of the island" (Murie, 1934: 42). He rejected the island's coyotes – he was also an expert on coyotes – as an ineffective predator for moose populations.

Wolves indeed proved effective. But the carnivores were not planted there – perhaps the Great Depression and World War II had something to do with that – but instead found Isle Royale on their own. By some accounts, the first pair of wolves walked to the island in 1949 over the ice from Ontario, and the four quiet decades spent by the moose population of Isle Royale without a natural predator would come to an end (Line, 1996).

National Park Service biologist James E. Cole was sent to Isle Royale from his regular assignment at Rocky Mountain National Park in Colorado to do a wolf count in 1952. He and park ranger Roger Hakala located four wolves during the winter. Another census four years later found "at least" 22 wolves; by 1957, Cole may have seen the final coyote track on the island – wolves had crowded them out (Cole, 1957).

In 1958, other researchers began a close study of the predator–prey relationship between wolves and moose that continues into the twenty-first century. The scientist credited with instigating the project was Durwood Allen, a professor at Purdue University in Indiana. Isle Royale had been set aside as a national park in 1940, ending moose hunting. Allen, as an undergraduate zoology student at the University of Michigan in 1930, had seen Hastings' film on the moose of Isle Royale and was excited when two large male wolf carcasses, taken as bounty, showed up at the university's museum. Twenty-one years later, working for the U.S. Fish and Wildlife Service in Washington, D.C., Allen returned to considering wolves, moose, and Isle Royale when a photograph of a wolf track landed on his desk. "[T]his event introduced the greatest of all experiments in predator–prey relations . . . I was especially interested in islands as research areas, where the animals you are counting and studying do not wander away," he wrote (Allen, 1979: XVII).

There were no federal funds for an Isle Royale study in the early 1950s, so Allen's work did not get underway until after his arrival at Purdue in 1954. He ended up leading the study from its inception in 1957 until 1976. His long list of graduate students came to include some of the world's leading wolf researchers, including L. David Mech and Rolf O. Peterson. The Isle Royale wolves were to become the most famous in America through television specials and newspaper and magazine articles (Figure 16.1). In fact, Peterson came to the project after seeing Allen on TV (Steinhart, 1995).

Figure 16.1 A lone moose holds off a wolf pack in the early 1960s on Isle Royale. After 5 minutes, the pack backed off.

Source: National Park Service

How did the first wolf or wolves arrive? Allen and his team concluded that the population they were observing came from Minnesota or Ontario on an ice bridge between 1948 and 1950. A confounding variable was the introduction of four trapped wolves in 1952, brought from the Detroit Zoo by a writer and wolf advocate named Lee Smits. After DNA testing on Isle Royale animals was introduced in 1990, the presence of the stocked wolves – even though two were shot and killed shortly thereafter and one was soon trapped and returned to the mainland – would prove controversial.

Mech, whose Ph.D. dissertation on the wolves and moose was published in 1966, determined that woodland caribou had occupied an ecological niche now partially filled by the moose, and Canadian lynx had been present on the island prior to the early 1930s (Mech, 1966). In three years of observations, including extensive aerial surveys, he concluded that the largest wolf pack on the island – 15 or 16 animals – killed an average of one moose every three days.

The mid-1960s in the United States were an ebb, in terms of population, for the wolves. State bounties were ending, to be certain, but almost too late to assist the carnivore population. Minnesota was down to about 350 wolves by 1965, while northern Wisconsin and the Upper Peninsula of Michigan had another 50, plus the 20 to 25 on Isle Royale.

Perhaps because of their small numbers, changing environmental attitudes, or other factors entirely, the wolves began to fall back into favor. With so few animals left in the continental United States, they could hardly be blamed by hunters or farmers for taking game or livestock in large numbers.

The scientists on Isle Royale and elsewhere spoke favorably of the wolf in public presentations, books, articles, and interviews in the popular press. Michigan gave the wolf protection under the state's endangered species law in 1965; in 1973, the wolf was included in the first federal Endangered Species Act. Animals listed in the ESA needed a recovery plan, and the wolf was no exception – except that the wolves on Isle Royale were written out of the plan because of their isolation (U.S. Fish and Wildlife Service, 2011). Nevertheless, shooting an ESA-listed gray wolf in the Lower 48 states could result in a $50,000 fine and a year in jail.

Perhaps nothing represents charismatic megafauna in the eyes of the public more strongly than a featured role in a children's book, and by 1968 the wolves and moose of Isle Royale were main characters in a book called *The Big Island*. Author Julian May told the story of a large, hungry moose that saw the island across the water in the distance and swam for it. Others followed, and still more came across later on an ice bridge. Drama ensued when overpopulation buffets the animals, and even humans cannot save them. Along comes the heroes, a wolf pack that is trapped on an ice floe and conveniently drifts to the island. And perhaps because it was a children's book – or especially biologically accurate – the wolf pack feeds only on the old, weak, and sick moose, balancing the population, so that the "wolves and the moose and other animals all live together. They all have plenty to eat" (May, 1968: 31). Those that were not eaten, the reader presumes, live happily ever after.

Rehabilitation of the wolf's reputation continued apace. In the 1970s, scientists formed working groups to study and protect the animal, including the International Union for the Conservation of Nature and Natural Resources (which became the World Conservation Union). Environmental groups such as the Audubon Society, the National Wildlife Federation, and the World Wildlife Fund also picked up the wolf's banner (Mech and Boitani, 2010). By the 1990s, public opinion surveys showed that citizens' attitudes toward the wolf supported protection (Mech and Boitani, 2010). In Ely, Minnesota, a zoo-museum-learning space called the International Wolf Center became a popular tourist destination. Under Mech's direction, a nonprofit was formed in 1989, and a modern facility was completed in 1993 and expanded in 1998 with the help of a grant from the state.

After being observed for so many seasons in relatively close quarters, perhaps it was inevitable that some of the predators would receive names. The two most famous were Big Jim and Old Gray Guy. Big Jim was the lone survivor of the

Detroit Zoo stocking, a 41-kilogram (90-pound) male that was raised in captivity – from one day old – and released at 15 months. As a budding charismatic megafauna, Big Jim sat for television cameras in Detroit and a New York magazine photographer before his release. Cole thought that a lone wolf he spotted over the years, including as late as 1957, was Big Jim, and his legend grew (Allen, 1979). But as a lone wolf, his genetic contributions as a stud were likely limited.

There were hopes, soon realized, for gene-mixing when a wolf quickly named Old Gray Guy (after his unusual light coloring) wandered to the island in 1997. He started a pack that grew to 10 animals, and his progeny composed more than half the animals on the island by 2006 (Bakalar, 2011). But, possibly because all the animals on the island except Old Gray Guy were descended from the same female – and possibly because genetic rescue simply does not work – inbreeding problems did not disappear.

Perhaps inevitably, the wolves became noted for a skeletal deformity. Canine parvovirus – possibly carried by dogs accompanying park visitors – has been a problem for the wolves. Moose escaped the inbreeding problem but often fell victim to starvation after over-browsing, or to a brain parasite abetted by a warming climate.

Thus Murie's sought-for ecological balance on Isle Royale, as represented in the children's book, did not last long if it ever existed. Instead, wolves and moose continued to survive in a peak-and-crash world that became extremely critical for the survival of both species on the island by the second decade of the twenty-first century.

The moose population appeared to peak in 1995 at about 2,450. The highest wolf count was 50 in 1980. By 2016, 500 moose were left, while the wolf population was down to two. Genetic rescue for the wolves was considered and rejected in 2014, although that decision was revisited in 2016 and is under consideration again at the time of this writing. The stocking proposal was controversial because opponents believed that some wolves may have descended from Big Jim rather than from wild wolves that wandered over from the mainland in the 1940s.

It has not been lost on policymakers that while the wolf population on Isle Royale has suffered, its recovery on the Lower 48 mainland states has been a success (Wydeven et al., 2009). Wolves have long been a cultural touchstone, both favorably and as the bad guy in fairy tales, but moose have not endured those reputational ups and downs. Now both animals on Isle Royale are under stress from a more recent challenge: climate change. As winters warm, the chances of an ice bridge–walking wolf or moose coming to the gene pool's rescue-lessens. Meanwhile, ticks and other often-deadly parasites can benefit from a warming planet.

Conclusion

In this way, the rise and fall of moose and wolves of Isle Royale represent the theme of climate change that other chapters in this volume touch on or examine more directly. In Part III, climate change is expressly addressed in the examination of toxicants and the water quality of the Great Lakes and the relationship that Native Americans have with black ash threatened by the invasive emerald ash borer. In terms of public policy, in Part IV the interconnected issues of water policy over the decades and the presence of invasive species – and threats to endangered ones – are discussed. Additional factors in citizen engagement and resource management contribute to the debate.

In Part I, habitat, conservation and restoration, dams, and irrigation are naturally connected in the critical controversy over water use, while artificial reefs and reef restoration add an element that is not often considered. While moose and wolves fight for survival on Isle Royale, Lake Huron turtles and their habitat challenges point to another important marker hundreds of miles to the east. The lessons learned from press coverage of Martha, the last known passenger pigeon in the world, should not be lost on policymakers, as described in Chapter 6. In Chapter 7, the wolf's recovery as it relates to another prey species, white-tailed deer, is presented in more detail. The movement of wolves and a charismatic megafauna not discussed in depth in other chapters, the black bear, in the northeastern Great Lakes gets important attention. Finally, Chapter 15 highlights the essential role that Indigenous peoples of the U.S. and Canada can play in tackling the conservation and environmental problems that confront all of us in the Great Lakes Basin.

Scientists from many agencies, research institutes, and universities across two countries have devoted their careers to studying environmental problems in the Great Lakes Basin. This volume brings together some of the most recent research in the field, while organizing the material in a way that allows the reader to make connections among the issues. The importance of their work is obvious; the need for continued attention, funding, and new ideas should be equally obvious. As scientists, public policy experts, lawmakers, tribes, and citizens of all stripes who care about the future of North America's greatest inland waters, the need has never been more urgent nor the task more daunting.

Note

1 Later, DNA evidence indicated that perhaps some moose were stocked by a Minnesota sportsmen's club for hunting purposes in the early 1900s. Locals report moose swimming to the island as well.

References

Allen, D. L. (1979). *The Wolves of Minong*. Boston, MA: Houghton Mifflin.

Bakalar, N. (2011, April 5). Wolf crosses the Lake Superior ice to become leader of the pack. *New York Times*.

Cole, J. E. (1957). *Isle Royale Wolf Investigations, Winter of 1956–57*. Unpublished manuscript, Isle Royale National Park files, Special Report, p. 42.

Grooms, S. (2002). A brief history of wolf research. *International Wolf, 21*(2): 9.

Hickie, P. F. (1943). *Michigan Moose*. Lansing, MI: Michigan Department of Conservation, Game Division.

Line, L. (1996, March 19). In long-running wolf-moose drama, wolves recover from disaster. *New York Times*.

May, J. (1968). *The Big Island*. Chicago, IL: Follett Publishing.

Mech, L. D. (1966). *The Wolves of Isle Royale*. Washington, DC: U.S. Government Printing Office.

Mech, L. D., and Boitani, L. (eds.). (2010). *Wolves: Behavior, Ecology and Conservation*. Chicago: University of Chicago Press.

Murie, A. (1934). *The Moose of Isle Royale*. Ann Arbor, MI. University of Michigan Press.

Steinhart, P. (1995). *The Company of Wolves*. New York: Vintage.

U.S. Fish and Wildlife Service. (2011). *Gray Wolf Recovery in Minnesota, Wisconsin and Michigan*. Available at www.fws.gov/midwest/wolf/aboutwolves/pdf/R3wolfrec.pdf

Wydeven, A. P., Van Deelen, T. R., and Heske, E. (eds.). (2009). *Recovery of the Gray Wolves in the Great Lakes Region of the United States*. New York: Springer.

Index

Note: Page numbers in *italics* indicate figures. Page numbers in **bold** indicate tables.

For Product Safety Concerns and Information please contact our
EU representative GPSR@taylorandfrancis.com Taylor & Francis
Verlag GmbH, Kaufingerstraße 24, 80331 München, Germany